DEDICATED TO THE MEMORY OF

PAT WALLER

DIRECTOR EMERITA OF THE
UNIVERSITY OF MICHIGAN TRANSPORTATION RESEARCH INSTITUTE

LIFELONG ADVOCATE FOR TRUCK DRIVER
OCCUPATIONAL SAFETY AND HEALTH

August 15, 2003

Cover photographs courtesy of Jim West / jimwestphoto.com.

Truck Driver Occupational Safety and Health

2003 Conference Report and Selective Literature Review[*]

Gregory M. Saltzman[†] and Michael H. Belzer[‡]

Revised February 8, 2007

DEPARTMENT OF HEALTH AND HUMAN SERVICES
Centers for Disease Control and Prevention
National Institute for Occupational Safety and Health

---

[*]The Wayne State University Truck Driver Occupational Safety and Health Conference was held April 24-25, 2003, at the Detroit Metro Airport Doubletree Hotel, Detroit, MI.

[†]E. Maynard Aris Professor and Chair, Department of Economics and Management, Albion College, 611 East Porter Street, Albion, MI 49224. Adjunct Research Scientist, Institute of Labor and Industrial Relations, University of Michigan. gsaltzman@albion.edu

[‡]Associate Professor of Industrial Relations, Wayne State University, Department of Interdisciplinary Studies, College of Liberal Arts and Sciences, Detroit, MI 48202. Adjunct Associate Research Scientist, Institute of Labor and Industrial Relations, University of Michigan. *Michael.H.Belzer@wayne.ed*

Disclaimer

Sponsorship of the Truck Driver Occupational Health Conference and these proceedings by the National Institute for Occupational Safety and Health (NIOSH) does not constitute endorsement of the views expressed or recommendations for the use of any commercial product, commodity, or service mentioned. The opinions and conclusions expressed in the presentations and report are those of the authors and not necessarily those of NIOSH. All conference presenters were given the opportunity to review and correct statements attributed to them within this report.

Recommendations are not final statements of NIOSH policy or of any agency or individual involved. They are intended to be used in advancing the knowledge needed for improving worker safety and health.

This document is in the public domain and may be freely copied or reprinted.

Ordering Information

To receive documents or other information about occupational safety and health topics, contact NIOSH at

NIOSH—Publications Dissemination
4676 Columbia Parkway
Cincinnati, OH 45226–1998

Telephone: 1–800–35–NIOSH (1–800–356–4674)
Outside the U.S.: 513–533–8328
Fax: 513–533–8573
E-mail: *niosh-publications@cdc.gov*

or visit the NIOSH Web site at *www.cdc.gov/niosh*

DHHS (NIOSH) Publication No. 2007–120

Foreword

In 2003, NIOSH co-sponsored a conference that brought together researchers from around the world to discuss the safety and health of commercial truck drivers. NIOSH recognizes that these workers merit attention due to the difficult and dangerous nature of their trade. Truck drivers have an unusually high rate of occupational injury, and one of the highest rates of on-the-job fatality. NIOSH is actively working to improve the safety and health of truck drivers. Current Institute projects will increase our understanding of cause-specific mortality among owner-operator truck drivers, the health effects of diesel exhaust particles, and the influence of work organization on truck driver fatigue.

Reducing occupational injury and illness among truck drivers is assisted by a coordinated effort, and this conference was an important step towards establishing a national research agenda. The following report and accompanying CD-ROM share the information, insight, and research of the professionals who participated in the conference. Together they provide an overview of the trucking industry, summarize the current state of knowledge regarding truck driver safety and health, and document the topics for future research suggested by the conference participants. NIOSH hopes that these proceedings will be valuable to researchers, industry representatives, policymakers, and the public.

John Howard, M.D.
Director, National Institute for
 Occupational Safety and Health
Centers for Disease Control and Prevention

Contents

Foreword	iii
Abbreviations	vii
Acknowledgements	ix
Executive Summary	x
Introduction	1
Keynote: The Need for a Research Agenda on Truck Driver Occupational Safety and Health	3
Introduction to Trucking: Operations, Labor Markets, and Occupational Safety and Health	6
Epidemiology, Surveillance, and Measurement	23
Ergonomics, Job Injuries, and Exposure	32
Labor Market, Employment Relations, and Personnel Management	39
Fatigue and Truck Driver Safety and Health	46
Causes of Driver Fatigue and Sleep Deprivation	57
Health Consequences of Driver Work Schedules	60
Health Consequences of Job Stress	62
Social and Behavioral Consequences of Employment as a Truck Driver	65
Developing a Research Agenda on Truck Driver Occupational Safety and Health	69
References	81
Appendix 1: Conference Participants and Speaker Profiles	96
Appendix 2: Agenda	114

The conference also is on the web at
http://www.ilir.umich.edu/TIBP/truckdriverOSH

Tables

Table 1. Occupational Injury/Illness of Commercial Motor Vehicle (CMV) Operators, 2004 1
Table 2. Driving Maximum Hours, Old Rules 19
Table 3. Working Maximum Hours, Old Rules 19
Table 4. Driving Maximum Hours, New Rules, Carriers Operating Six Days a Week 19
Table 5. Driving Maximum Hours, New Rules Carriers Operating Seven Days a Week 20
Table 6. Working Maximum Hours, New Rules Carriers Operating Six Days a Week 20
Table 7. Working Maximum Hours, New Rules Carriers Operating Seven Days a Week 20
Table 8. Permanent Disability Claims for Unionized LTL Carriers 34
Table 9. French Truck Driver Work Hours 42
Table 10. Research Needs: Data 74
Table 11. Research Needs: Economics and Industrial Organization 75
Table 12. Research Needs: Assessment 76
Table 13. Research Needs: Interventions and Countermeasures 78
Table 14. Research Needs: Dissemination and Education 80

Abbreviations

ATA	American Trucking Associations
BAC	Blood alcohol concentration
BLS	Bureau of Labor Statistics
BMCS	Bureau of Motor Carrier Safety (predecessor to FMCSA)
BTS	Bureau of Transportation Statistics
CDLIS	Commercial Driver License Information System
CFOI	Census of Fatal Occupational Injuries
CPSC	Consumer Product Safety Commission
DHHS	U.S. Department of Health and Human Services
DOT	U.S. Department of Transportation
EEC	European Economic Community (predecessor of European Union)
EEG	Electroencephalogram
EOBR	Electronic on-board recorder
EU	European Union
FACE	Fatality Assessment and Control Evaluation
FARS	Fatality Analysis Reporting System
FMCSA	Federal Motor Carrier Safety Administration
GH	Growth hormone
HOS	Hours-of-service regulations
IBT	International Brotherhood of Teamsters
ICC	Interstate Commerce Commission (abolished 1995)
IHD	Ischemic heart disease
INRETS	French National Institute for Transport and Safety Research
JAMA	Journal of the American Medical Association
LTCCS	Large Truck Crash Causation Study
LTL	Less than truckload
MCMIS	Motor Carrier Management Information System
NAFTA	North American Free Trade Agreement
NCHS	National Center for Health Statistics
NEISS	National Electronic Injury Surveillance System
NHAMCS	National Hospital Ambulatory Medical Care Survey
NHTSA	National Highway Traffic Safety Administration
NIOSH	National Institute for Occupational Safety and Health
NOMS	National Occupational Mortality Surveillance System
NTOF	National Traumatic Occupational Fatality Surveillance System
OOIDA	Owner-Operator Independent Drivers Association
OSA	Obstructive sleep apnea
OSHA	Occupational Safety and Health Administration
OSPAT	Occupational Safety Performance Assessment Test
PVT	Psychomotor vigilance test
REM	Rapid eye movement sleep
SENSOR	Sentinel Event Notification System for Occupational Risk
SOII	Survey of Occupational Illnesses and Injuries

SWS	Slow-wave sleep
TIBP	Trucking Industry Benchmarking Program
TIFA	Trucks Involved in Fatal Accidents
TL	Truckload
UMTIP	University of Michigan Trucking Industry Program
UMTRI	University of Michigan Transportation Research Institute
VIUS	Vehicle Inventory and Use Survey
3PL	Third party logistics

Acknowledgements

The Wayne State University Truck Driver Occupational Safety and Health Conference was organized by Michael H. Belzer and sponsored by the National Institute for Occupational Safety and Health (NIOSH), the Owner-Operator Independent Drivers Association (OOIDA), the International Brotherhood of Teamsters (IBT), the Trucking Industry Program and the Trucking Industry Benchmarking Program (both partly sponsored by the Alfred P. Sloan Foundation as part of the Industry Centers Program), with in-kind resources provided by the Douglas Fraser Center for Workplace Studies in the College of Urban, Labor, and Metropolitan Affairs at Wayne State University.

The authors thank the eight anonymous referees for their comments on the draft of this report, and thank Jan Birdsey of NIOSH for help in preparing the final version.

Executive Summary

In April 2003, an international group of researchers convened in Detroit to discuss the occupational safety and health of commercial motor vehicle drivers. This conference was unusual because it focused on driver well-being, rather than general highway safety and transportation issues. Truck drivers merit special attention not only because of their large numbers—approximately 2.8 million in the U.S.—but also because they face extraordinary risk of on-the-job injury and death. In 2004, U.S. truck drivers were 7 times more likely to die on the job, and 2.5 times more likely to suffer an occupational injury or illness, than was the average worker.

The meeting was sponsored by the National Institute for Occupational Safety and Health, the Owner-Operator Independent Drivers Association, the International Brotherhood of Teamsters, and the Trucking Industry Program and the Trucking Industry Benchmarking Program at Wayne State University. The following report provides a selective review of the relevant literature, summarizes the conference presentations, incorporates the comments made by many of the participants, and outlines some topics needing further research. The accompanying CD-ROM contains the conference PowerPoint presentations, conference handouts, supporting papers, and reports. Information in this Executive Summary was presented or discussed at the conference, and relevant literature references are provided within the report.

Introduction to Trucking

The first plenary session provided an overview of the trucking industry and government regulations. The U.S. trucking industry is essentially divided into three main segments: government entities, private carriers, and for-hire motor carriers. Government entities such as the U.S. Postal Service operate truck fleets for their own use. Private carriers (wholesale, service, or industrial companies) may sell transportation services to others, although they primarily haul their own freight. For-hire motor carriers receive all of their freight from shippers. The for-hire trucking industry is commonly grouped either by shipment size (parcel, less-than-truckload [LTL], and truckload [TL]), or by geographic scope (local, regional, inter-regional, national, or international) and type of commodity or operation.

The structure of the U.S. trucking industry changed radically soon after deregulation in 1980. Formerly regulated common carriers that had assumed health care and pension costs found it difficult to compete with new non-union TL carriers that had no pension or health care liabilities and little capital overhead. Trucking has shifted from being predominantly unionized before deregulation in 1980 to less than 15% unionized today. Driver wages (adjusted for inflation) have fallen 30% since 1980.

The Federal Motor Carrier Safety Administration issued the latest hours-of-service rules for commercial truck drivers in 2005.[1] The new rules increase the minimum daily off-duty time, yet allow drivers to spend more total time working and more time behind the wheel than the previous regulations. For solo drivers, the mandatory break period after a shift has increased from 8 to 10 hours, offering greater opportunity for a driver to obtain restorative sleep, and all work must be completed within 14 hours of the driver's start time, making it more difficult to falsify log books. However, the new rules also increase the maximum allowable daily driving time from 10 hours to 11 hours, and they allow drivers to work as much as 84 hours in a seven-day period. Drivers were previously limited to 70 hours of work time within any seven-day interval, but a new "restart" provision allocates a fresh set of work hours to a driver after 34 hours of continuous off-duty time (theoretically, two sleep periods). Thus, a driver can now complete 70 hours of duty time early in the 5th day of the week, take 34 hours off, and work another 14 hours in day 7, for a total of 84 hours.

U.S. drivers are allowed to work more hours than their counterparts do in either the European Union or Australia's eastern states. While the daily work schedule allowed in Australia's eastern states (14 hours of work and 12 hours driving in a 24-hour period) is similar to the U.S. hours of service rules, the Australian drivers are limited to 72 hours of working time in a seven-day period, 12 hours less than the U.S. "restart" provision allows. Drivers in the European Union (EU) are allowed even fewer hours; the latest EU rules will eventually limit all truck drivers to an average of 48 hours per week (over a four-month period), and a maximum of 60 hours per week. Currently, all EU drivers are limited to 9 hours of driving per day.

Long working hours, irregular shift and day-off patterns, and many consecutive nights away from home and family can make truck driving an arduous career. The overtime pay provisions of the Fair Labor Standards Act, which require time-and-one-half pay for work hours in excess of 40 per week, do not apply to truck drivers. Drivers are often paid by the mile, and they are not compensated for the time they spend waiting at the docks or performing tasks such as loading and unloading, which make up approximately 25% of their total work time. The American Trucking Associations cite the combination of undesirable working conditions and low wages as reasons for the 121% driver turnover rate in the large truckload sector for 2005.

Truck Driver Occupational Safety and Health

Several sessions at the conference outlined existing surveillance efforts and highlighted a variety of safety and health topics. Participants agreed that while several existing data systems collect detailed information on highway crashes and on-the-job fatalities, significant data gaps remain, especially in the areas of fatigue, occupational stress and violence, and chronic disease. With new hours-of-service rules now in place, complete and timely data on truck driver health and safety is essential to characterize health risks, identify emerging problems, and assess interventions.

Chemical and ergonomic hazards

Truck drivers face a variety of chemical exposures. Drivers who deliver unhardened concrete are exposed to chromium and alkaline substances, putting them at risk for both allergic skin reactions and chemical burns. Drivers of gasoline tanker trucks often experience acute headaches, dizziness, or nausea after exposure to vapors released during gasoline transfers. Diesel exhaust has been linked to lung cancer and allergic inflammation, and exposure can be substantial at loading docks and truck stops. Increasing the allowable working hours for drivers increases the risk that their exposures to hazardous substances will exceed recommended limits.

Ergonomic risks vary widely and can include loading and unloading heavy cargo (beverage delivery drivers handle approximately 36,000 pounds each day), awkward postures, and working in tight spaces such as the drum of a ready-mix concrete truck. The effects of these hazards are borne out by the high number of injuries reported by drivers both in the U.S. and abroad.

Fatigue and highway safety

Many of the conference participants considered fatigue to be a critical problem for drivers and the trucking industry. Safe vehicle operation requires sustained vigilance, excellent judgment, and quick reactions, particularly during heavy traffic or poor driving conditions. Fatigue impairs all of these abilities, endangering not only truck drivers, but also other motorists who share the road with them. A common proximate cause of fatigue among truck drivers is partial sleep restriction, which in turn causes sleep debt and semi-chronic sleepiness. One study reported that drivers slept an average of only 4.78 hours per day, while those who worked on a steady night schedule averaged only 3.83 hours of sleep per day.

Sleep restriction, in turn, stems both from medical problems such as sleep apnea and from employment conditions in trucking. Long work hours cut into time available for sleep, and work-related stress makes it hard for drivers to sleep even when time is available. Shift work or an irregular work schedule forces many drivers to sleep during the day, in opposition to natural circadian rhythms, and most team drivers probably experience this irregular schedule. Sleeper berths do not provide optimal sleeping conditions, and sleep likely is fragmented for solo drivers and team drivers alike. Some drivers find it difficult to sleep in a moving vehicle, especially if they do not trust the driving ability of their partner, or if the vehicle is making numerous stops. Likewise, solo drivers' sleep may be fragmented while awaiting notification of the availability of their next load.

Even moderate levels of sleep deprivation (17-24 hours of wakefulness) can cause neurobehavioral impairment equivalent to a blood alcohol level (BAC) of 0.05 to 0.10. This is considered an unsafe level of functioning for truck drivers; commercial drivers may have their licenses revoked if they operate a heavy truck or haul hazardous cargo with a BAC of 0.04 or above. Further compounding the problem, a driver may not be aware of his or her level of impairment for two reasons: (1) over several days of sleep restriction, habituation may make sleepiness feel normal to the driver, and (2) "wake state instability" causes individuals with moderate sleep loss to perform optimally most of the time, but not reliably. Thus, while sleep-deprived individuals do not necessarily experience any immediate impairment of neurobehavioral function, lapses can occur when sustained cognitive accuracy and speed are critical.

The inability to determine their level of impairment may explain why some drivers are willing to risk driving while fatigued. In the U.S., 25% of long-distance truck drivers reported falling asleep at the wheel in the past year, while 47% reported having fallen asleep at the wheel sometime during their driving career. Drivers were more likely to report falling asleep at the wheel if they split their off-duty periods, or if they worked a demanding schedule (10 or more hours of consecutive driving, or less than 8 hours of off-duty time per day). Similarly, Australian truck drivers reporting 6 or fewer hours of sleep prior to a trip were significantly more likely report a hazardous event related to fatigue during that trip, such as nodding off while driving. According to conference participants, driver fatigue was a factor in 6% to 49% of the highway truck crashes they studied.

Other health effects of driver work hours

In addition to increasing the risk of being in a crash, the long and irregular work hours of many drivers can have other adverse health effects. Sleep debt is associated with impaired glucose metabolism, abnormal cortisol (stress hormone) regulation, altered growth hormone (GH) profiles, altered autonomic function (elevated sympathovagal balance), and impaired immune function. Disruptions in the endocrine system may contribute to health problems such as obesity, diabetes, and hypertension. Long-haul drivers who work many hours per day and do not have 24-hour work/rest cycles may suffer from desynchronized internal circadian rhythms. Working over 40 hours per week can double the risk of acute Helicobacter pylori infection (associated with peptic ulcers), even controlling for age, sex, and marital status. Irregular hours and night work raise the risk of being hospitalized for ischemic heart disease (IHD), and professional drivers are at greater risk of IHD if they work long hours.

The demanding work schedule of truck drivers also makes it more difficult for them to obtain quality health care. Forty-seven percent of long-distance truck drivers surveyed lacked a regular health care provider. Fifty-six percent of drivers found it difficult to make a healthcare appointment when at home due to their work schedule, and 62% said they had failed to seek out needed health care when on the road working.

Stress

Truck drivers experience stress from many sources. Long working hours, night work, or spending extended periods on the road away from friends and family can isolate drivers and leave them too exhausted to nourish their relationships. Pressure to stay on schedule even when road conditions are bad or they are fatigued can strain drivers' nerves. Delivering or picking up loads can be taxing – drivers are often required to wait in their trucks for long and unpredictable periods of time; they may be denied opportunities for food, water, and restroom facilities; and they may be treated disrespectfully by shipping and receiving personnel. Owner-operators are often under intense financial pressure, finding it difficult to make the required loan or lease payments on their truck due to low compensation rates.

Workplace violence

Truck drivers can be victims of either physical violence or verbal abuse. Among 300 Australian truck drivers surveyed, 30% had been victims of verbal abuse, 21% had been victims of "road rage," 10% had been threatened, and 1% had been assaulted. At freight forwarding yards, verbal abuse and threats were closely linked with economic pressures in nearly all incidents. Loading delays, drivers cutting in line, and mistakes by forklift drivers fueled tensions and led to the violent behaviors. U.S. beer delivery drivers are robbery targets, since they may carry $1,000 to $3,000 in cash by the end of the day. Threat of such abuse and violence can increase psychological stress and use of maladaptive coping mechanisms.

Driver Compensation and Safety

Driver compensation is inversely associated with both employee turnover and crash risk. Three empirical studies on the relationship between truck driver compensation and safety found a strong and significant relationship: more pay to the driver resulted in substantially fewer crashes. In a study of J.B. Hunt drivers, researchers found a 10% higher base mileage rate was associated with a 34% lower probability of a crash. In addition, a 10% increase in driver pay was associated with a 6% lower crash probability, giving an overall pay-rate effect of 1:4. In a cross-sectional study of 102 truckload carriers, researchers found that for every 10% increase in average total compensation for drivers, the carrier would experience a 9.2% lower crash rate. In a driver survey study, researchers found that drivers earning a 10% higher mileage rate had an 18.7% lower probability of having a crash during the reporting year, while drivers given 10% more paid days off had a 6.3% lower probability of having a crash during that same year.

Higher compensation may reduce crash risk by enabling carriers to recruit and retain better-qualified drivers. For example, unionized LTL carriers provide good pay (typically $60-70,000 per year for drivers), pensions, and health insurance benefits; drivers at these companies tend to have long job tenure, sometimes staying at the same company for 30 to 40 years. In addition, higher compensation might reduce the driver's economic incentive to exceed legal and safe driving limits while increasing his or her desire to maintain a good driving record.

Research Needs

The conference participants cited a need for additional research on a wide variety of topics. Many discussions revolved around fatigue; how to measure it and compare it to other types of impairment such as alcohol intoxication; how to help drivers recognize when they become dangerously sleepy; and whether fatigue can be effectively managed through prescription drugs, light therapy, or changes to work organization. Fatigue was clearly a topic of concern, partially because this was an area of expertise for many of the participants, and partially because it appears to be a pervasive problem among drivers.

Other suggestions included:

- ⇒ Explore the association between work characteristics and driver health
- ⇒ Explore the relationship between driver compensation and safety, and quantify the society-wide benefit/cost ratio of increasing driver compensation
- ⇒ Explore ways of increasing truck driver access to health services, including continuity of care needed for effective diagnosis and treatment
- ⇒ Determine whether drivers' health-related behavior can be improved by increasing their awareness of safety and health risks
- ⇒ Compare the health risks and work characteristics of U.S. drivers to those of other countries
- ⇒ Determine the relationship between sleep debt and endocrine function
- ⇒ Study long-term effects of demanding work schedules and sleep debt

In evaluating the potential health effects of the revised hours-of-service rules for trucking, the Federal Motor Carrier Safety Administration stated that due to lack of existing data on commercial truck driving, "…the Agency had to evaluate and weigh information from different fields and adapt it to a trucking environment."[1] This statement makes it clear that more research specifically on commercial truck drivers is needed. Ideally, a research agenda addressing the most pressing issues will be developed in consultation with truck drivers, trucking companies, government agencies, and other interested parties. Such an agenda would provide direction to the safety and health community and ensure that concerns are addressed in a coordinated manner.

Introduction

Many truck drivers suffer from occupational safety and health problems. In 2004, truck drivers accounted for 15% of the fatal occupational injuries in the U.S. (873 of 5,764).[2] Similarly, in 2004, they accounted for 8% of the nonfatal occupational injuries and illnesses involving days away from work (100,730 of 1,259,320),[3] even though they comprise only 2% of the workforce[4] (see Table 1). In part, the substantial number of drivers dying on the job or suffering occupational injuries or illnesses involving lost work time reflects the large number of truck drivers in the U.S.: somewhat over 1.5 million drivers of heavy trucks or tractor-trailers and almost 940,000 drivers of light or delivery trucks in 2004, not counting self-employed drivers.[4] Nevertheless, truck driving was among the occupations with the highest rates of fatal occupational injuries in 2004.[5]

Table 1. Occupational Injury/Illness of Commercial Motor Vehicle (CMV) Operators, 2004[2,3]

	Fatalities	Percent	Injury/illness	Percent
Total—all occupations	5,764	100%	1,259,320	100%
Bus drivers	19	0.33%	3,330	0.26%
Truck drivers, heavy, and tractor trailer	779	13.51%	63,570	5.05%
Truck drivers, light, and delivery	94	1.63%	37,160	2.95%
Total truck drivers	873	15.15%	100,730	8.00%
Total CMV drivers	892	15.48%	104,060	8.26%

In order to focus attention on the seriousness of truck driver occupational safety and health problems and develop an agenda for future research, a conference on this topic convened on April 24-25, 2003, in Detroit, Michigan. The conference, which was organized and hosted by Professor Michael H. Belzer of Wayne State University, brought together over five dozen experts from government, universities, employers, industry associations, and labor from the United States, Europe, and Australia (see Appendix 1). This report provides a selective review of the relevant literature, summarizes the 30 presentations at the conference, incorporates the comments made by many of the conference participants, and outlines some topics needing further research (see Appendix 2 for the conference agenda, and the accompanying CD-ROM or *http://www.ilir.umich.edu/TIBP/truckdriverOSH* for the PowerPoint presentations, conference handouts, and supporting papers and reports). It also presents selected citations to the relevant scholarly literature.

The conference was organized around research and problem areas important to occupational safety and health researchers and practitioners, as well as those areas of concern identified by experts in trucking industry and labor issues. Trucking industry representatives attended and participated, including those working in the union and nonunion sectors of the industry. The Owner-Operator Independent Drivers Association (OOIDA) and the International Brotherhood of Teamsters (IBT, to which we refer hereafter as the "Teamsters") represented owner-drivers as well as employee drivers, ensuring that the research and industry communities were listening to each other. Professor Belzer attempted to include in the conference experts with a wide range of relevant specialties, but it was beyond the scope of this conference to develop a consensus statement of research priorities. All statements in this report not attributed to a conference presentation or a comment made by a conference participant are the responsibility of the authors of this report, Gregory M. Saltzman and Michael H. Belzer. The opinions expressed in this report are those of the conference participants or of Professors Saltzman and Belzer, and not necessarily those of the National Institute for Occupational Safety and Health.

Keynote: The Need for a Research Agenda on Truck Driver Occupational Safety and Health

Dr. Patricia F. Waller
Director Emerita
University of Michigan Transportation Research Institute

Dr. Patricia F. Waller was unable to attend due to the illness that took her life on August 15, 2003, before this conference report was complete. Her prepared remarks, which follow, stress the importance of a research effort on behalf of truck driver health and safety. As one of the first women and non-engineers involved in the Transportation Research Board nearly 40 years ago, she observed the lack of interest in this subject and began a lifelong effort to address commercial driver wellness issues. She notes the scarcity of good data about truck crashes when she first began to study the subject in the 1970s. She also saw a need for further investigation into the reasons why truck drivers tend to develop chronic diseases such as diabetes at relatively early ages. She commended the conference organizer and sponsors for taking these steps to develop a research agenda on truck driver occupational safety and health.

> First, I want to say how delighted I am that this meeting is being held. It has long been needed.
>
> When I first became involved with heavy truck safety in the 1970s, there was essentially no attention paid to the driver, except to blame him when anything went wrong. Research on truck safety was funded either by the government, for purposes of regulation, or by the industry, for purposes of profit. The driver was essentially ignored. Yet it was the driver who was held responsible for meeting competing demands, often incompatible demands, and making the whole system work.
>
> There was essentially one database on truck crashes, the BMCS file. This file consisted of crash reports that were filed with the federal government, voluntarily, by the motor carriers. The validity of the information was more or less taken for granted.
>
> In the 1970s, the Federal Motor Carrier Safety Standards called for drivers to use available seat belts. But when I looked at truck crashes occurring in North Carolina, using the state crash file, and compared it with the BMCS crashes occurring in North Carolina, I found that the reports in the BMCS file said that 85 percent of the drivers were using their belts at the time of the crash. The North Carolina police reports found that only 15 percent of the truck drivers were using their belts.
>
> But there were much more serious problems. A study by Ken Campbell at University of Michigan, examining only fatal truck crashes nationwide, found that for trucks coming under ICC authorization, 19 percent of the

fatal crashes were not in the BMCS file. For trucks not under ICC authorization, the proportion was 49 percent. And of course fatal crashes are much more likely to be reported.

Nevertheless, the BMCS file was what was used for most truck crash analyses. A very few independent studies were conducted and found very different results. When discrepancies were brought up, the response was, "But we have 11 studies here (based on BMCS data) that show thus and so, and there are only two studies, based on independently collected data, showing something different." So of course the 11 studies won out.

Good data were very hard to come by. And there was a general attitude that if you did not have data showing there was a problem, that proved there wasn't a problem. And almost no one was concerned about the plight of the driver. OOIDA was still a fledging organization working to help truck drivers become successful small businessmen (or women). In more recent years OOIDA has published major articles on health promotion topics focused on truck drivers. But there was still no body of literature examining the issue of truck driver health and well-being.

In the late 1970s, the US DOT conducted public hearings on truck safety. Originally scheduled to last a day and a half, it soon became evident there would need to be at least one more day. Drivers came from as far away as Alaska for their five minutes at the microphone. They came in every size and shape imaginable, but they told very similar stories. They described being forced to take overloads, being forced to take trucks that did not have good brakes or were in some other way dangerous. They were forced to meet schedules that could not be met without violating safety regulations. And of course the whole issue of loading and unloading and hours of service led to dangerous situations.

Meanwhile, the Bureau of Labor Statistics regularly published a list of the occupations that accounted for the most days lost under workers' compensation for illness or injury. Truck drivers hold the top spot and have for some time. But these data were coming from an entirely different federal agency, and there was no communication or coordination with DOT.

At the 1995 National Summit on Truck Safety, there was a physician who worked with the Teamster's Union. He said that when he was first approached about serving in this capacity, he knew nothing about truck drivers or trucking. However, he got involved, and he was soon struck by the fact that truck drivers were developing chronic illnesses, such as diabetes, at much younger ages than was true for the general population. Still, there was no effort being made to delve into what was accounting for this phenomenon.

Part of the reason for the failure to examine the truck driver's plight was that there was no one really qualified to do it, that is, there was no one who had a real understanding of the truck driver and also had the skills necessary to conduct the necessary research. And of course there was no obvious source of funding. Truck drivers had no one championing their concerns, with adequate funding to investigate the problem.

When I met Mike Belzer and learned of his background, I thought, "At last, here is someone who could make things happen in this area." I want to express my delight that all of you consider this an important topic and are here to develop a research agenda. It is long overdue, but it is exciting to see it get under way. I will look forward to the results of this meeting. And thank you, Mike, for inviting me to come.

Introduction to Trucking: Operations, Labor Markets, and Occupational Safety and Health

The conference presentation by Dr. Stephen Burks (University of Minnesota, Morris) provided an overview of trucking services in the U.S. "For-hire" motor carriers –carriers that offer their services for sale to the market, and are understood to comprise the trucking industry – provide only a portion of trucking services. Private carriers ("not-for-hire" retail, wholesale, service, or industrial companies that operate truck fleets primarily for their own use) or government (U.S. Postal Service, federal, state, and municipal governments) provide the rest. Private carriers have more heavy freight vehicles than for-hire carriers do, but for-hire carriers account for more total miles per year. The average for-hire truck is used more intensively: for-hire trucks operate an average of 82,000 miles per year, compared to 32,000 miles per year for private trucks.[*] The for-hire trucking industry is commonly divided into segments either by shipment size (parcel; less-than-truckload, or LTL; and truckload, or TL, including specialized haulers), or by geographic scope (local, regional, inter-regional, national, or international) and type of commodity or operation. Different technologies and specialized production equipment are used for each segment. Barriers to entry are substantial in parcel, moderate in LTL, and minimal in TL, except among the specialized haulers, some of whom have very specialized equipment and customer networks. While most truck drivers appear to operate similar equipment and use similar general skills, many require specialized training, including dry and liquid bulk haulers, car haulers, operators of refrigerated trucks, dump truck drivers, and others.

Dr. Burks did not have time in his presentation to describe the nature of these barriers to entry or the consequences of trucking market structure; therefore, the authors have added the following explanation. Economists recognize two broad classes of barriers to entry: institutional (regulatory) restrictions and economic barriers. Larger transportation firms have a competitive advantage over smaller firms if there are economies of route density or economies of scope. Economies of route density lower costs per unit for truckload firms when there are many shipments per week for a particular origin-destination pair, and for the less-than-truckload and packages sectors when there are many shipments per day between city pairs or within a local pickup and delivery network.[†] Economies of scope lower costs per unit when a single firm produces two outputs jointly, rather than having independent firms produce them separately.[6] Economies of scope can be spatial, if adding routes enables a firm to increase capacity utilization for the firm's vehicles;[7] they

---

[*] The Vehicle Inventory and Use Survey (VIUS), used here to analyze the trucking function and also used to provide a denominator for analyses of motor carrier safety and productivity, was terminated by Census in 2006.

[†] The distinction in transportation economics between economies of density (more trips per week on a particular route) and economies of scale (larger firm size) was made by Douglas W. Caves, Laurits R. Christensen, and Michael W. Tretheway, "Economies of Density versus Economies of Scale: Why Trunk and Local Service Airline Costs Differ," *The RAND Journal of Economics*, Vol. 15, No. 4, Winter 1984, pp. 471-489.

could also reflect cost savings if a single firm jointly produced LTL and parcel services. In a sense, all intercity LTL firms produce a joint product: local pickup and delivery, and intercity truckload services, because operationally the intercity move is a truckload type of move. There may be economies of density[8] and economies of scope in LTL and parcel delivery. Stigler proposed a survivor technique to infer the existence of economies of scale: if firms below a certain size tend to lose market share in an unregulated market, then one can infer that these firms were smaller than minimum efficient scale.[9] A survivor-technique study found that market share of LTL firms with less than 2 million ton-miles of freight per year was substantially lower in 1993 (after deregulation) than in 1975 (before deregulation), suggesting that smaller LTL firms have a competitive disadvantage,[10] although this study did not distinguish between the impact of economies of density and economies of scope. Recent consolidation in LTL and parcel delivery provides further evidence that there may be economies of density or scope in LTL and parcel.

In December of 2003, Yellow Transportation, a national LTL carrier, merged with Roadway Express to become by far the largest LTL company in America, and in May of 2005, Yellow-Roadway acquired USF – a national network of regional and inter-regional carriers – to further consolidate its dominance in LTL. Federal Express, originally a package carrier, purchased Viking (a west coast regional) and American Freightways (an east-of-the-Rocky-Mountains interregional) to establish FedEx Freight, an LTL carrier with national reach. In 2005, FedEx bought Watkins Motor Lines, a trans-national LTL carrier, to further extend its national capacity. In 2005, United Parcel Service acquired Overnite, one of the nation's largest interregional carriers, purchasing it from Union Pacific, which had hauled the trans-national freight on the railroad. These consolidations have transformed the entire LTL sector, creating an entirely new competitive environment.

In contrast, the TL sector has perhaps hundreds of thousands of small firms. Consistent with the economic model of perfect competition, a single small TL firm has almost no power to influence the price of the trucking services it sells; it is at the mercy of market forces beyond its control. But the growth of large and sophisticated TL carriers such as J.B. Hunt and Schneider National suggests that some segments of TL may have economies of density or economies of scope. Large TL firms often develop subsidiary operations providing third-party logistics services ("3PLs") to shippers and consignees in an effort to take advantage of economies of scope to attract freight and fill up their network (which gives them economies of density). It also provides additional services (two outputs, in economists' parlance) in an effort to increase their firm's value to their customers and giving the customer an opportunity for "one-stop shopping."

Market concentration in trucking is similar in the U.S. and Europe, but Dr. Dieter Plehwe (Social Science Research Center Berlin and, at the time of the conference, Yale University) stated that trucking in the European Union (EU) has experienced paradoxical change. While the package express sector has consolidated, as have some of the large trucking companies that span the entire EU, small- and medium-sized firms have actually become even smaller, fueling a growing gap between large and small firms.*

Many TL firms consist of a single owner-operator who owns one truck, arranges for freight, operates under his or her own authority (hence the term "owner-operator") and drives the truck. Data from the University of Michigan Trucking Industry Program suggest that approximately 85% of all owner-drivers lease both their truck and themselves to a larger motor carrier, under whose authority they operate. More recent data collected by the Wayne State University Trucking Industry Benchmarking Program suggest that this fraction may have dropped to as low as 65%, although these data may be biased toward more independent operators since they were collected on line from OOIDA members in a very sophisticated survey.[13] While most of these individuals either own their trucks outright or obtain financing from a bank or financing company, some of them – many with very little personal capital – lease their trucks from a firm loosely affiliated with the carrier to which they are leased. This quasi-employment/debtor relationship may put them into debt peonage, in which they incur large financial obligations to pay for the capital cost of leasing or buying their truck and then find it difficult or impossible to generate enough revenues to meet these obligations, much less have something left over to pay themselves wages for their driving labor. The leasing company can declare the "owner"-driver in default if he or she is unable to meet the lease obligations, whereupon it repossesses the truck and leases it to someone else. Like the coal miners in Tennessee Ernie Ford's song, "Sixteen Tons," they owe their souls to the company store.

Owner-operators often have had little training in financial management of business, much less the incredibly complex business of interstate trucking, with its maze of licensing and taxation regimes and intense competition. Dr. Burks characterized the situation of many owner-operators as a failure of rational choice; a stream of new owner-operators continually enters the industry, even though the average owner-operator appears to make less money than employee drivers, use up their business capital in a relatively short time, and go out of business at a high rate. John Siebert (Owner-Operator Independent Drivers Association) noted that the highly competitive market structure in TL shipping services causes productivity improvements to drive down shipping prices, rather than to raise the earnings of owner-operators above those of employee drivers (who do not undertake the risk of purchasing and maintaining capital, and operating a business).† Another

---

* See also Plehwe, Dieter, "National Trajectories, International Competition, and Transnational Governance in Europe," in G. Morgan, P. H. Kristensen, and R. Whitley (eds.), *The Multinational Firm: Organizing across Institutional and National Boundaries* (Oxford: Oxford University Press, 2001), pp. 281–305 and Dr. Plehwe's conference presentation on the accompanying CD-ROM.

† For a recent analysis of owner-operator cost-of-operations and earnings, see Belzer, Michael H. (2006), 'OOIDA 2003-2004 Cost of Operations Survey: Report of Results', (Detroit, MI: Wayne State University). This on-line survey of 421 owner-drivers shows that owner-operators who have no employees and drive their own trucks earn an average net operating profit per mile of 36.4 cents, of which 6.8 cents is wages they pay to themselves. See www.ilir.umich.edu/TIBP. Available from the author by request.

conference participant characterized many owner-operators as bogus entrepreneurs, saying that owner-operators who always work for the same companies are covert employees. Indeed, the unclear standards governing the status of what some other nations call "dependent owner-operators" continue to be an issue.[11] Although the National Labor Relations Board was supposed to have clarified the matter,[12] court judgments in recent lawsuits in California and elsewhere have found that dependent owner-operators (such as those driving package delivery trucks for FedEx) do not meet the tests of independence sufficient to distinguish them from employees.

Dr. Michael Belzer (Wayne State University and the University of Michigan Institute of Labor and Industrial Relations) proposed a framework for understanding the nature of the economic and operational problems faced by truck drivers. His presentation was based on his trucking industry, truck driver, and motor carrier safety research, including research reported in his book, *Sweatshops on Wheels*.[13] The structure of the trucking industry changed radically soon after deregulation in 1980, as formerly regulated common carriers "burdened" with health care and pension costs – as well as with hitherto required terminal overhead – failed to compete with new-entrant truckload carriers that were non-union, had no pension or health care liabilities, and little capital overhead.[14] Trucking shifted from being predominantly union-represented before deregulation in 1980 to less than 15% unionized today.[15] This has caused trucking firms to compete on the basis of wage costs, and just as carriers have little bargaining power to influence rates, individual drivers have little bargaining leverage or market power to raise their wages.

Drivers work long hours; for example, intercity drivers work an average of 65 hours per week, which is 5 hours more per week than legally allowed.[*] However, about 25% of their work time is unpaid. Ten percent of all over-the-road drivers work 94 hours or more per week. Drivers' individual decisions to work illegally long hours contribute to their problem by supplying excessive labor to the marketplace in what becomes a self-perpetuating sweatshop environment. Drivers' work schedules also are irregular, both in terms of days of the week worked and time of day worked, and this irregularity likely contributes to both safety and health problems.

Truck drivers' wages are low: the University of Michigan Trucking Industry Program's Truck Driver Survey in 1997 found that nonunion truckload drivers earned the equivalent

---

[*] Under regulations in effect at the time of the conference and until January 4, 2004, drivers technically could work to complete tasks they were performing at the time they ran out of hours, but they could not drive after they have worked 60 hours in a seven-day period or 70 hours in an eight-day period. The regulations that became effective in January 2004 have almost the same weekly limit rules except that drivers who use up their weekly hours can invoke the 34-hour restart provision, allowing them to work legally up to 84 hours per week. Regulatory enforcement, however, has been a long-standing issue because it is based primarily on driver self-report. Drivers historically have falsified their Records of Duty Status ("log books") to exceed weekly limits, and both public and private surveys have demonstrated that the top 25% exceed these weekly limits by 25% and the top 10% exceed them by 50% – too many hours to be explained by non-driving work beyond the daily limit.

of a job with overtime premium pay that had a base wage of just $8.17 per hour.[16] See also Belman and Monaco (2001).[17] Current Population Survey data also indicate that driver earnings are low; drivers who work up to the legal limit earn about $7 per hour, and those who work more hours than the legal limit earn $6.20 per hour overall.[18] Driver wages have fallen 30% in real (inflation-adjusted) terms since 1980 due to the elimination of barriers to entry, deregulation of prices for trucking services, the dramatic increase in interfirm competition, and the drastic decline of unionization in trucking.[19] Drivers often live in their trucks and are away from home for extended periods. As a result, employee turnover is high, particularly in the truckload sector – typically 100% per year in the non-union truckload sector (where wages are substantially lower than in unionized sectors). In recent years, the average turnover rate has been steady at about 130%.

Labor economist John Budd has argued that employers' demand for efficiency should be balanced with employees' demand for equity and voice. He bases his argument on a business ethics paradigm; while he allows that people have different ethical models, he points out that people must have an ethical framework of some kind. The balance among the three objectives of the employment relationship – efficiency, equity, and voice – can be analyzed using the ethics of utilitarianism (based on Jeremy Bentham and John Stuart Mill, stressing efficiency), the ethics of liberty (based on John Locke, stressing property rights), the ethics of duty (based on Immanuel Kant, stressing respect for human dignity and the Golden Rule), the ethics of virtue (based on Aristotle, stressing moral character), the ethics of justice (based on John Rawls, stressing personal liberty, equal opportunity, and concern for the poor), and the ethics of care (based on Carol Gilligan's notion of nurture).[20] Businesses and citizens make their decisions based on these ethical frameworks. In Belzer's view, as more trucks become sweatshops on wheels, policy-makers have given too much attention to efficiency and too little attention to equity and voice in the supply chain. Although truckers have come to understand that low truck driver compensation underlies the labor supply problems the industry faces, the shipping community continues to take advantage of cheap freight based on low compensation and low trucking industry profits.

Trucking industry competition, Dr. Belzer said, has adversely affected highway safety. Low wages have forced the industry to hire drivers with undesirable characteristics, created perverse incentives that motivate drivers to cheat on their logs and motivate carriers to turn a blind eye to such cheating, and this has made it difficult to retain experienced qualified drivers.[21] The increased use of just-in-time delivery has put additional pressure on drivers to meet tight schedules. Dr. Belzer argued that reducing driver hours (especially unpaid labor time) and increasing driver compensation could improve highway safety. Charging shippers and consignees for delay time would reduce unproductive use of driver time, thereby minimizing lost production associated with reductions in driver work hours, and reducing drivers' incentive to log unpaid time as non-work time and exceed legally prescribed hours-of-service limits. It would mitigate the problem that economists call "moral hazard," or living on "somebody else's money:" shippers would absorb the cost of delays for which they are responsible.*

These pressures on drivers to work longer hours and run more miles for less money puts drivers in a no-win situation. Dr. Gregory Saltzman (Albion College and the University of Michigan Institute of Labor and Industrial Relations) discussed the need for federal regulation of truck driver work hours to keep competition within socially acceptable bounds. He noted that federal hours-of-service (HOS) regulations have limited work hours of interstate truck and bus drivers since 1938. Violations of HOS regulations, however, are widespread. Furthermore, long and irregular driver hours adversely affect highway safety and driver health. He recommended that the federal government enforce existing HOS regulations with electronic on-board recorders (EOBRs), similar to those required in the European Union, noting that EOBRs would reduce the tendency of trucking companies to base their competitive advantage on their willingness to violate the law. He also recommended that HOS regulations be amended to guarantee a weekly day of rest and to foster 24-hour work/rest cycles.

In an idealized economic model, Dr. Saltzman noted, there would be no need for government HOS regulations. If long work hours have an adverse effect on drivers' health, family life, and sense of well being, then drivers will demand higher hourly pay for jobs that entail long hours. Trucking companies and shippers will be willing to pay the cost of these compensating wage differentials only if the benefit to them of having drivers work long hours is large enough to justify the cost. In short, market forces reduce driver hours whenever the benefits from reduced hours exceed the costs.

Dr. Saltzman added, however, that economists recognize two circumstances that would justify government regulation notwithstanding the above argument. First, the distribution of income may not be ethically acceptable. Truck drivers work longer than average hours and face a high risk of occupational fatality, for lower than average hourly pay.

---

* The Federal Motor Carrier Safety Administration announced new Hours of Service rules on the day of this conference. One part of the new rules may help to solve this problem, albeit indirectly. Although many experts have considered this interpretation illegal, under the HOS rules existing at the time of the conference as they were interpreted by the Federal Highway Administration's Office of Motor Carriers (FMCSA's predecessor), drivers had been allowed to log as much time off duty during any "tour of duty" as they wished. Since road drivers almost always are paid only "when the wheels are turning," they typically would log loading, unloading, and waiting time as "off duty." Under the new rules, truck drivers "may not drive beyond the 14th hour after coming on-duty, following 10 hours off duty." Since "breaks do not extend on-duty time" under the new rule, either driver productivity will decline or shippers and consignees will become much more efficient in the use of drivers' time. While they may still log time off during their 14-hour tour-of-duty, they will not be able to extend their work day. See final rule: Department of Transportation, Federal Motor Carrier Safety Administration, 49 CFR Parts 385, 390, and 395, "Hours of Service of Drivers; Driver Rest and Sleep for Safe Operations," *Federal Register*, Vol. 68, No. 81, Monday, April 28, 2003, pages 22456–22517.

Second, market failures prevent an unregulated market for truck driver labor from achieving economic efficiency. Truck drivers' long work hours creates what is known as an externality – a situation where some of the costs or benefits of a decision spill over to third parties not involved in making that decision. Long work hours for truck drivers endanger the safety of other highway users, and some of the damage caused by highway collisions is either not fully compensated or is compensated through insurance policies that are not fully experience-rated. As a result of this externality, an unregulated free market provides less than the efficient level of highway safety. Similarly, the unregulated market creates a health-related externality because some of the costs of unhealthy working conditions for drivers spill over to third parties. These third parties include insurance companies (because trucking companies generally are too small to have fully experience-rated health insurance plans), subsequent employers or taxpayers (because health effects may become manifest years later, when the driver is employed elsewhere or is covered by Medicare), and health providers (because of uncompensated care for current or former drivers without health insurance).

Another class of market failures includes inadequate or asymmetric information and cognitive problems in processing data. The long latent period for occupational illnesses such as diabetes and cardiovascular disease prevents drivers from recognizing the importance of changing their employment conditions until much of the health damage has already been done. Furthermore, as Akerlof and Dickens noted, cognitive dissonance often causes workers to underestimate the health risks they face in the workplace, as allowing themselves to recognize these risks would be too stressful.[22] Because of these information and cognitive problems, labor markets do not force trucking companies and, ultimately shippers, to pay drivers higher wages to compensate them fully for assuming health risks. Similarly, decisions by managers may be distorted when some costs are not explicit. Since nonunion long-haul and regional drivers typically are paid only by the mile, with no separate pay for non-driving work, trucking companies and shippers perceive driver waiting and loading time as free; they therefore fail to re-engineer inefficient aspects of their operations. Finally, economic theory does not assert that existing regulations continue to be efficient. Hours-of-service regulations designed at a time of regulated trucking rates and maintained in an era of strong union protection for drivers, high production costs for EOBRs, and limited scientific knowledge of adverse health and safety effects of sleep deprivation, may no longer be appropriate today.

One potential area for future research would be to assess the impact of a small business training program on the safety records of owner-operators. Are owner-operators who understand their costs less likely to face intense financial pressures that lead them to drive when they should be sleeping? Another potential area for future research would be to determine whether the self-employed drivers in the U.S., like those in Australia, are more likely to have safety problems. Are self-employed drivers in the U.S. more likely to have nonfatal or fatal crashes?

In response to this presentation, Dr. Michael Quinlan noted that self-employed workers in Australia have higher fatality rates in mining as well as in trucking because there is less regulation for self-employed workers and because self-employed workers are less likely to be unionized. He said that unionization of long-haul truck drivers in Australia had fallen dramatically as employee drivers became self-employed, even though they have a protected right to unionize (unlike similarly employed U.S. drivers). Dr. Quinlan reported that the Transport Workers Union had pushed for a bill in the Australian federal Senate to set a minimum compensation rate per kilometer for owner-operators in order to improve highway safety and protect the interests of owner-operators; but the bill seemed unlikely to pass soon.

Dr. Michael Quinlan (University of New South Wales) presented information centered on the impact of commercial arrangements and competitive forces on occupational safety and health in Australian long-haul trucking. During 2000 and 2001, he prepared a report on this topic for the Motor Accidents Authority of New South Wales[23] (document included on accompanying CD-ROM and available at *http://www.ilir.umich.edu/TIBP/truckdriverOSH*). Intense competition among carriers and drivers, he said, causes instability and insecurity. Commercial pressures stemming from low rates adversely affect safety, Dr. Quinlan continued, because irresponsible and desperate carriers undercut responsible ones. The result is evasion of occupational safety and health regulations and inadequate maintenance. It is also difficult for regulators to obtain cooperation from drivers. Logbooks for hours are called "cheat sheets" in Australia; long hours, speeding, and drug use are common among drivers. In cases of serious road incidents, drivers are reluctant to give evidence in court that will cause them to lose their jobs.

Dr. Quinlan gave particular emphasis to the role of shippers and consignees, who control rates, scheduling, and service because they generally are larger and more powerful than motor carriers, and as the source of carriers' business have powerful market leverage. Shippers and consignees have little incentive to improve driver safety and health because the costs of driver occupational safety and health problems are borne by others. The subcontracting chain obscures their legal responsibility. According to Dr. Quinlan, lawyers inform shippers that the less they know about the operations of the motor carriers they hire, the less they are liable for safety violations. Dr. Quinlan cited a need to establish a chain of responsibility, in which "all parties to commercial arrangement for supplying a service or good have legal duties, not just those who carry out the actual task." He noted that shipper liability laws have been enacted in Saskatchewan and Manitoba and that one is being considered in California. His 2001 report proposed a requirement that shippers and consignees (as well as drivers) sign documents attesting to trip routes, departure times, and arrival times to provide an evidentiary trail regarding driver hours. It also called for demurrage clauses (so that shippers and consignees minimize waiting time) and for a ban on arrival time bonuses and penalties (so that drivers will be under less time pressure). Dr. Quinlan argued that making shippers and consignees legally responsible would change health and safety outcomes by shifting responsibility to those who actually dominate the freight transportation process.

Other conference participants agreed with Dr. Quinlan's emphasis on shippers. Ms. Katharine Newman argued that one cannot design effective mitigation strategies for dangerously long driver work hours without involving shippers. One participant suggested that shippers would come to the table only when they are confronted with a mandate regarding the imposition of driver hours, while others emphasized the importance of establishing a chain of responsibility. Dr. Dieter Plehwe reported that in Germany, the shipper is liable for violations of regulations regarding driver work hours. This stops shippers from pushing prices for trucking services down to such low levels that HOS violations are inevitable. An implication of the conference discussion is the following ethical question for consumers who benefit from low shipping costs: what level of risk are people willing to accept in trade for cheaper goods? In the case of trucking, consumers share much of the truck drivers' safety risks because the public highway is the truck drivers' assembly line.

Trucking firms have fixed costs: freight terminals, network development, and route-optimization systems for package delivery and LTL firms; network development and route-optimization systems for TL firms; and the capital cost of trucks for all trucking firms. Intensive capital utilization reduces average costs for trucking firms. Dr. Plehwe commented that trucking firms need to operate their trucks intensively in order to make money. To accomplish this, carriers may demand that drivers run more miles and carry more freight. Another strategy is to have more than one driver per truck. Drivers who team with another driver can operate 24 hours a day, seven days a week (though not in Central Europe). Trucks can also operate around the clock by "slip-seating" drivers – rotating drivers in the same truck, either as a relay for over-the-road operations, or with sequential shifts for local work. Many drivers, however, do not like to share trucks with other drivers, and especially do not like to run team. Mr. Ron Jager, an owner-operator, noted that owner-operators tend to avoid co-drivers because they do not want to put their lives into somebody else's hands. Mike Belzer, a former over-the-road driver, concurred from the perspective of an employee driver.

Dr. Dieter Plehwe (Social Science Research Center Berlin and, at the time of the conference, Yale University) gave a presentation on trucking regulation in the European Union (EU). He noted a tension between two perspectives. There is widespread support for improving driver safety and health and for reducing air pollution from trucks. Nevertheless, free market advocates oppose economic regulation, and small motor carriers cannot afford increased costs for labor, fuel, and road use.

In the 1990s, Dr. Plehwe continued, the EU gradually deregulated rail transport, but it rapidly deregulated trucking. Cross-border liberalization allowed more Eastern European truck drivers to work in Western Europe, a trend accelerated by the economic collapse of Eastern Europe. Yet large disparities remain within Europe in working time and internalization of external cost (i.e., social and environmental harmonization is slow). Dr. Plehwe noted that these disparities are larger among the countries of the EU than they are within in the U.S. and perhaps similar to those within NAFTA.

Dr. Plehwe explained that there are two categories of EU rules. European Union "directives" ("soft law") require national legislatures to adopt laws meeting EU minimum standards. European Union "regulations" ("hard law") take effect immediately, without any need for action by national legislatures. In the 1990s, there was a substantial increase in EU regulations related to transport.

Some of the EU transport rules were aimed at creation of a single European market. EU Regulation 92/881 of 1992 made it easier for EU carriers to operate across national borders within the EU. Regulation 93/3118 of 1993 changed cabotage rules[*]; it allowed carriers from one EU state to handle shipments whose points of origin and delivery were both in other EU states. After these regulations were adopted, trucking competition intensified within the EU, driving down shipping prices and driver wages.

Other EU rules relate to driving time and total working time. Regulation 3820/85 of 1985 limited drivers, including owner-operators, to an average of 9 hours of driving per day. Directive 93/104/EC regulating work hours for most EU workers did not apply to the transport sector. Employer/union negotiations in rail reached an agreement on working hours, but these negotiations failed in road transport. In March 2002, the European Parliament adopted Directive 2002/15/EC, which required national legislatures to limit working hours for truck drivers to an average of 48 hours per week (averaged over a 4-month period), and a maximum of 60 hours per week. These limits will apply to employee drivers no later than March 2005, and to self-employed drivers no later than March 2009.[24]

Dr. Plehwe said that limits on driving time are not effectively enforced anywhere. He predicted, however, that enforcement will be assisted by the EU requirement of second-generation tachographs. These will be digital and will record more data than the driving hours and driving speed recorded by first-generation tachographs.

One important topic for future research is to assess the effectiveness of the EU's second-generation tachographs in improving truck driver occupational safety and health. These devices (called electronic on board recorders, or EOBRs, in FMCSA's May 2000 hours-of-service proposal) can enforce limits on driving time for single-driver operations, though overall limits on work time might still be exceeded because the devices will not record waiting time or other non-driving work time. As noted in a later section, Dr. Patrick Hamelin reported that non-driving work time accounted for about 30% of total driver work time, so the inability to record non-driving work time is a major limitation. The effectiveness of EOBRs is further compromised in team operations where one driver operates the vehicle and another driver sleeps. Even with a coded and password-protected electronic card, two drivers may collude to use each other's card and password to split the driving however they wish. Privacy concerns may prevent the adoption of a more secure system that uses biometrics or face recognition technology to provide electronic tracking and logging of truck driver activity, though a similar biometric control has been specified

---

[*] Trans-national carriers engage in "cabotage" when they haul freight from their home country to another country, and then haul domestic freight (point to point pickup and delivery) on movements within that destination country. For example, a Bulgarian motor carrier that hauls freight into France would be performing cabotage if it delivered freight to Paris and picked up freight in Paris destined for Marseilles.

as part of the Transportation Worker Identification Card that soon will be required in the U.S. by Department of Homeland Security regulations. Nevertheless, even without biometrics or face recognition technology, EOBRs are substantially harder to falsify than are the paper work logs now used in the U.S. American policy makers may be able to learn from EU experience with second-generation tachographs; if the European experience is favorable, then field trials could be conducted in the U.S. in which motor carriers or owner-operators are provided incentives to use EOBRs in their trucks. Dr. William Rogers predicted that EOBRs would have a greater impact on the TL sector than on the LTL sector as drivers in the TL sector are less likely to be in compliance with HOS regulations.

April 2003 Revisions in U.S. Hours-of-Service Rules

On April 24, 2003 (the first day of the conference), FMCSA announced the first significant changes since 1962 in hours-of-service regulations covering interstate truck drivers.[*] On the evening of April 24, these new HOS rules were discussed by a panel including Mr. David Snyder, Vice President and General Counsel for the American Insurance Association (who was in Washington for the FMCSA announcement and participated in the conference by speaker phone); Mr. Robert Rothstein, General Counsel for the American Moving and Storage Association; and Mr. Ron Jager, an owner-operator.[†] The new rules, which took effect on January 4, 2004, apply to truck drivers but not to interstate bus drivers (who remain under the old rules). Some aspects of these new rules are applicable to certain other drivers, while some drivers continue to be exempted from portions of the old and new rules. Partially exempt industrial operations include oil field operations, ground water well drilling operations, construction materials and equipment, and utility service vehicles; agricultural operations are entirely exempt insofar as they do not exceed a 100-mile radius of operations and operate during planting and harvesting seasons as determined by the states. The new rules make several changes:

---

[*] The Circuit Court of Appeals struck down the new HOS regulations in July 2004 on the grounds that FMSCA "neglected to consider a statutorily mandated factor - the impact of the rule on the health of drivers." See Public Citizen v. Federal Motor Carrier Safety Administration, 374 F.3d 1209 at 1216 (D.C. Circuit, 2004). The motor carrier industry especially did not want to return to the old regulation, having spent a great deal of money restructuring their operations and re-educating their employees to comply with the new regulation. After a period of legal and political wrangling, Congress passed a law in September 2004, giving FMCSA one year to develop new regulations and an analysis that would comply with the court's ruling. FMCSA issued a new rule in September 2005, modifying the HOS rules originally issued in April 2003, by requiring drivers using the sleeper berth provision to take at least 8 consecutive hours in the sleeper berth, with the modification effective October 1, 2005; it also issued a new analysis addressing health issues. See
http://www.fmcsa.dot.gov/rules-regulations/topics/hos/hos-2005.htm, available online on June 27, 2006, and included on the accompanying CD-ROM and at *http://www.ilir.umich.edu/TIBP/truckdriverOSH*.

[†] Note: this discussion is included at this point in the conference report, rather than at the conclusion of the first day's report, because it pertains to the issues of regulation and operation.

⇒ They allow truck drivers to drive for 11 hours in each work/rest cycle, rather than in the previous rule.
⇒ They prohibit an individual from driving after the 14th hour after having come on duty for the current work shift (i.e. drivers cannot extend their work shift by taking breaks or by logging loading, unloading, or waiting time "off duty").
⇒ They require that each work/rest cycle include an off-duty period of 10 hours, rather than 8 in the previous rule.
⇒ For team drivers, the rules allow this 10-hour off-duty period to be divided into two separate segments, provided that no segment is less than two hours long.[*]
⇒ They allow drivers who have been on duty for 60 hours within 7 consecutive days (for six-day-per-week operations) or 70 hours within 8 consecutive days (for seven-day operations) to resume work after an off-duty "restart" period of 34 consecutive hours, rather than having to wait until the 7-day or 8-day period[†] is over.

States will have three years to change their regulations to make them consistent with the new HOS rules.

The new restart provision allows truck drivers to *drive* more hours per week. Under the new regulations, if a driver works for a company that operates six days per week, he or she can start at midnight on Sunday morning, and have 21-hour drive/rest cycles (11 hours of driving and 10 hours off duty). Thus, a driver can complete 60 hours of driving in a period of 110 hours, rotating his or her circadian clock backward three hours per day. The driver's first workweek, therefore, could last approximately 5 ½ days, spanning 110 hours. The 34-hour restart makes him or her available to drive again at midnight on Saturday morning with a fresh set of hours, for a theoretical maximum of 74 hours of driving in a seven-day period (see table 4).

The new provisions have particular importance for long-haul firms, which typically operate on a seven-day week (which invokes application of the eight-day rule). Drivers would run out of driving time early on their sixth day of work, allowing them to drive 70 hours in a 124-hour time span but be eligible to start fresh again as soon as 8 PM on the seventh day. They also would be able to drive as many as 74 hours in a seven-day period (see table 5). In all cases, it allows as much as a three-hour backward rotation of the wake-sleep cycle per day, though in practice the impact is unclear.

---

[*] The rules effective October 1, 2005, require a single eight-hour period for sleep, plus the additional two hours. The rules require the driver to remain in the berth for the entire eight hours, a requirement to which many drivers object.

[†] These seven- and eight-day periods constitute the truck driver's work week and this is common parlance in the trucking industry (the FMCSA refers to these periods as the "week" and the "workweek" in the regulation), but the 74- and 84-hour maximum to which we refer in subsequent paragraphs applies to the seven-calendar-day week.

If a driver has non-driving work duties (and most drivers do), then the reset provision would also allow more hours of total work per week. A driver on a 14/10 schedule (14 hours of work, including at least 3 hours of breaks or work other than driving) and working for a company that operates six days per week could run out of hours early on the fifth day (a 100-hour time span), but end up working as many as 84 hours during a seven-day week (see table 6).[*] A driver working for a firm that operates seven days per week – a more likely scenario – would be permitted to complete 70 hours of work in five days, running out of hours after a workweek spanning 110 hours. After the restart period on day six, the driver could begin with fresh hours, working a full fourteen-hour day on day seven (see table 7). This driver also can work 84 hours in seven days, or 20% more possible working hours than under the previous rules.

Tables 2–7 illustrate scenarios involving maximum working time under both old and new rules. For all tables, we assume that drivers take no breaks and work the maximum hours on all days. Realistically, the driver will take breaks and may well log waiting or non-driving labor time as off duty, extending the number of hours available during a week. This will not extend his or her daily hours, however, since they are limited by the 14-hour rule.

Indeed, the most significant effect of this rule will be felt on a daily basis. When a driver starts fresh with 14 hours to work (including 11 hours of driving time), he or she may not drive beyond the 14th hour. This means that delays experienced en-route (either from traffic or from customers who force the driver to wait to load or unload), whether logged on-duty or off-duty, will count against his or her overall daily work time available. In addition, drivers might work less than the available 14 hours, log off for 10 hours, and then start a new work shift. This will cause the driver's start time to vary, potentially rotating the driver's circadian schedule forward or backward. Both of these real-world examples may significantly reduce the maximum hours estimates above.

An important question is how the change in HOS regulations will affect productivity. While FMCSA estimates a 3.9% increase in long-haul productivity (primarily derived from the additional driving hour),[25] J.B. Hunt estimates a 2.6% drop in utilization and a 2.4% decline in on-time performance, using the 34-hour restart.[26] Schneider estimates productivity declines up to 20%, depending freight characteristics and length of haul productivity enhancement on the part of shippers and consignees.[27] It is important to distinguish between labor productivity (output per hour) and production (total output), however, as an increase in driver hours might raise production but cut labor productivity

---

[*] Note that even though drivers work for carriers that operate 6 days per week, these drivers can work 7 days per week; how a carrier not operating 7 days a week has drivers working 7 days a week is not explained. Although they may not exceed 60 hours of work per week, the 34-hour restart negates the effect of the rule because they can drive 74 hours per week and work 84 hours per week, which is a 37% increase in working time. While Table 4 shows the driver driving all available working hours taking a day off, this is arbitrary because normally driver schedules vary from this extreme with unpredictable affects on hours of work. Table 5 shows a driver for the same kind of carrier unambiguously working 7 days a week.

Table 2. Driving Maximum Hours, Old Rules

Day	Time of day	Hours	Totals
Day 1	12 am – 10 am	10	10
Day 1 – 2	6 pm – 4 am	10	20
Day 2	12 pm – 10 pm	10	30
Day 3	6 am – 4 pm	10	40
Day 4	12 am – 10 am	10	50
Day 5 – 6*	6 pm – 4 am	10	60
Day 6**	12 pm – 10 pm	10	70
Day 7	OFF DUTY	0	70

* For those on 7-day week, out of hours until midnight, start of Day 8
** For those on 8-day week, out of hours until midnight, sart of Day 9

Table 3. Working Maximum Hours, Old Rules

Day	Time of day	Hours	Totals
Day 1	12 am – 3 pm	15	15
Day 1 – 2	11pm – 2 pm	15	30
Day 2 – 3	10 pm – 1 pm	15	45
Day 3 – 4*	9 pm – 12 pm	15	60
Day 4 – 5**	8 pm – 6 am	10	70
Day 6	OFF DUTY	0	70
Day 7	OFF DUTY	0	70

* For those on 7-day week, out of hours until midnight, start of Day 8
** For those on 8-day week, out of hours until midnight, start of Day 9

Table 4. Driving Maximum Hours, New Rules, Carriers Operating Six Days a Week

Day	Time of day	Hours	Totals
Day 1	12 am – 11 am	11	11
Day 1 – 2	9 pm – 8 am	11	22
Day 2 – 3	6 pm – 5 am	11	33
Day 3 – 4	3 pm – 2 am	11	44
Day 4	12 pm – 11 pm	11	55
Day 5*	9 am – 2 pm	5	60
Day 6*	OFF DUTY	0	60
Day 7	12 am – 11 am	11	71
Day 7	9 pm – 12:00 am	3	74

Drivers on seven-day log
* 34-hour restart between 2 pm on Day 5 through 12 am, Day 7

Table 5. Driving Maximum Hours, New Rules
Carriers Operating Seven Days a Week

Day	Time of day	Hours	Totals
Day 1	12 am – 11 am	11	11
Day 1 – 2	9 pm – 8 am	11	22
Day 2 – 3	6 pm – 5 am	11	33
Day 3 – 4	3 pm – 2 am	11	44
Day 4	12 pm – 11 pm	11	55
Day 5	9 am – 8 pm	11	66
Day 6*	6 am – 10 am	4	70
Day 7*	8 pm – 12 am	4	74

Drivers on eight-day log
* 34-hour restart between 10 am on Day 6 through 8 pm, Day 7

Table 6. Working Maximum Hours, New Rules
Carriers Operating Six Days a Week

Day	Time of day	Hours	Totals
Day 1	12 am – 2 pm	14	14
Day 2	12 am – 2 pm	14	28
Day 3	12 am – 2 pm	14	42
Day 4	12 am – 2 pm	14	56
Day 5*	12 am – 4 am	4	60
Day 6–7*	2 pm – 4 am	14	74
Day 7	2 pm – 12 am	10	84

Drivers on seven-day log
*34-hour restart from 4 am on Day 5 to 2 pm on Day 6

Table 7. Working Maximum Hours, New Rules
Carriers Operating Seven Days a Week*

Day	Time of day	Hours	Totals
Day 1	12 am – 2 pm	14	14
Day 2	12 am – 2 pm	14	28
Day 3	12 am – 2 pm	14	42
Day 4	12 am – 2 pm	14	56
Day 5	12 am – 2 pm	14	70
Day 6*	OFF DUTY	0	70
Day 7	12 am – 2 pm	14	84

Drivers on eight-day log
* 34-hour restart goes from 2 pm on Day 5 to 12 am on Day 7

if fatigue diminishes the effectiveness of drivers. Further empirical research on the impact of HOS regulations on both productivity and production may be useful for policy makers.

David Snyder praised certain aspects of FMCSA's new HOS rules. He said that they followed the findings of sleep science by moving closer to 24-hour work/rest cycles (21 hours if driving time is maximized under the new rules, compared with the 18-hour work/rest cycles allowed under the 1962 amendments to the HOS rules), even if he would have preferred that they move farther in that direction. He supported the increase in off-duty periods to at least 10 hours in each work/rest cycle. Nevertheless, Mr. Snyder was disappointed that the new HOS rules did not address enforcement with EOBRs and that the 34-hour restart period allows drivers to work more hours per week than the old rules do. He also said that the new rules are no simpler than the old rules and thus, no easier to administer.

Mr. Rothstein reported that FMCSA had not run a field test of the new HOS rules prior to announcing them. He said it would have been preferable to have run such a field test to determine the effects of the new rules on driver sleep and operational requirements of motor carriers. Nevertheless, Mr. Rothstein generally supported the new rules.

Mr. Rothstein pointed out that the new HOS rules were simpler than the HOS proposal that FMCSA put forth in May 2000, which had different rules for each of five classes of drivers. He was pleased that the new rules did not require EOBRs. He said that the industry objected to the cost of EOBRs. He also noted that the American Trucking Associations had expressed concern that EOBRs would put truck owners and drivers at a disadvantage in crash litigation where a truck and a car collide; the truck's speed and the number of hours the truck driver had been driving would be recorded by an EOBR, but comparable information about the car would not be available.* Finally, Mr. Rothstein supported the provision in the new rules allowing the 10-hour off-duty period, if taken in the sleeper berth, to be broken into two separate segments.

Dr. Ann Williamson disagreed with Mr. Rothstein on the latter point. She argued that breaking the 10-hour off-duty period into two segments would reduce the amount of sleep that drivers get.† Dr. Williamson also expressed disappointment that the new HOS rules do not address the time of day that drivers work, because nighttime driving is more dangerous.

Mr. Jager noted that brokers with whom owner-operators often deal typically have more power than individual owner-operators. Mr. John Siebert of OOIDA concurred, noting that brokers frequently ask more trucks to come than the number of available loads. Mr.

---

* John Siebert, of OOIDA, later agreed with Mr. Rothstein. He said that owner-operators were very independent minded and therefore opposed EOBRs as intrusive.

† In the amended rules issued in September of 2005, the sleeper team provision was changed to require at least one break of at least eight hours.

Snyder noted that the amendments to HOS regulations proposed by FMCSA on April 24, 2003 put no new liability on brokers. They do not expand the chain of responsibility. Drivers continue to have sole legal responsibility for HOS violations even if the broker tenders them a load with unreasonably little time for delivery.

Dr. Drew Dawson expressed concern that the new HOS rules could result in drivers getting less sleep than the old rules did because the 34-hour restart period allows them to work more hours per week. Dr. Orfeu Buxton expressed concern about the three-hour backward rotation permissible under the new rules if a driver's work/rest cycle consisted of 11 hours of driving and 10 hours of rest. Sleep would then be expected to occur on a regular basis at a circadian phase unfavorable to good-quality, restorative sleep, Dr. Buxton said.

Mr. Scott Madar of the Teamsters said that younger drivers might like the new HOS rules because they can make more money, but older drivers might not because the 34-hour restart period may mean less time off at the end of the week.

FMCSA's May 2000 HOS proposal, one conference participant noted, was very widely disliked. The April 2003 HOS amendments, in contrast, have some supporters, even though others are disappointed. Many conference participants, however, agreed that an important goal for research was to make future HOS regulations more science-based.

Epidemiology, Surveillance, and Measurement

Several speakers at the April 2003 conference addressed issues of epidemiology, surveillance, and measurement. Transportation safety analysts traditionally have focused on highway crash data, driver fatigue, work hours, and driving speeds. Epidemiologists have taken a broader view, considering not only crash injuries, but also disease frequency and the relationship between risk factors and disease. Accurate measurement is essential to identification of driver occupational safety and health problems and assessment of the effectiveness of interventions to address these problems. A good surveillance system also includes development of standard metrics for evaluation.

Dr. Roger Rosa (National Institute for Occupational Safety and Health [NIOSH]) started the discussion of surveillance and measurement by defining surveillance as systematic data collection over a period of years—tracking, rather than taking a single snapshot. Dr. Rosa's remarks are summarized well by his conference report (also included on the accompanying CD-ROM and at *http://www.ilir.umich.edu/TIBP/truckdriverOSH*):

> Research on work-related sleep loss, sleepiness, or fatigue suggests that these factors may increase the risk of operational errors contributing to traumatic injuries or catastrophic incidents. In addition, chronic sleep loss, circadian rhythm disruption, or fatigue from demanding work schedules may contribute to stress-related health problems. Tracking these relationships and estimating their frequency of occurrence in the U.S. working population, however, has been difficult. Currently, there is little or no information in existing occupational safety and health surveillance systems to link injuries or illnesses with fatigued or sleep-deprived workers. Establishing those links requires the development of reliable indicators of sleep loss and fatigue risk that can be recorded systematically and efficiently, either through existing surveillance systems, or by establishing new forms of surveillance. Successful tracking of fatigue-related cases would help focus injury and illness prevention efforts.
>
> Laboratory studies of sleep loss and field studies of demanding work schedules point to some sleep loss and fatigue risk-factor indicators that might be developed for use in surveillance systems. Among the possibilities are: extended hours awake, reduced sleep, disrupted sleep, number of hours worked (e.g., long shifts, overtime), number of consecutive days worked, night or rotating shifts, unpredictable work times and on-call scheduling, high workloads, high environmental demands (e.g., heat, noise), reduced job efficiency, operational errors and critical incidents, degraded mood and feelings, physiological changes, and home and family demands. The extent to which any of these potential indicators can be integrated in surveillance systems depends on how easily and reliably they can be quantified, how frequently they might occur in a work situation, and how much they contribute to risk.

Several Federal surveillance systems and programs have been established to track work-related injuries, illnesses, deaths, and emerging conditions. There are two major systems for tracking fatal injuries: the National Traumatic Occupational Fatality Surveillance System (NTOF), maintained by NIOSH, and the newer Census of Fatal Occupational Injuries (CFOI), maintained by the U.S. Bureau of Labor Statistics (BLS). The NTOF has no indicators that might be associated with sleep loss or fatigue. The CFOI records the time of day and day of week of fatal injury. Some minimal information on night work as a contributing factor might be developed from those data. NIOSH also maintains the National Occupational Mortality Surveillance System (NOMS) to examine work-related deaths from illness. This system compiles data from death certificates obtained from the CDC National Center for Health Statistics (NCHS). There are no sleep loss or fatigue indicators in the NOMS.

Nonfatal illness and injuries rates are estimated in the Survey of Occupational Illnesses and Injuries (SOII), maintained by the BLS. SOII data are obtained from standardized injury or illness reports (OSHA 200 logs) submitted by private industry employers. Currently, there are no fatigue indicators in the SOII. An upcoming revision of the OSHA logs, however, will require employers to record the time of injury and the time the injured individual started work. Some indication of fatigue from consecutive hours worked might be inferred from those data.

Other more specialized surveillance programs could be modified to obtain information on sleep loss and fatigue. For example, the National Electronic Injury Surveillance System (NEISS), maintained by the Consumer Product Safety Commission (CPSC), and the National Hospital Ambulatory Medical Care Survey (NHAMCS), maintained by NCHS, monitor emergency department visits and already collect some work-related information. The CPSC developed NEISS to monitor injuries involving consumer products and to serve as a source for follow-up investigation of selected product-related injuries. NIOSH has an agreement with CPSC to collect work-related injury data in a sample of 69 hospitals in NEISS. Currently, there is little information on fatigue in the NEISS, but it may be possible to design follow-up investigations of selected cases to determine whether sleep loss or fatigue were contributing factors to the injury.

The Sentinel Event Notification System for Occupational Risk (SENSOR) is another specialized program using State-based collaborative agreements with NIOSH to target selected work-related conditions for prevention activities. Targeted conditions have included acute pesticide poisoning, asthma, carpal tunnel syndrome, lead poisoning, noise-induced hearing loss, amputations, silicosis, and youth occupational injury. Cases are identified through several sources, such as physician reporting, death certificates, hospital discharge data, and workers' compensation records. Examples of prevention activities used in SENSOR programs are information dissemination, education, worksite consultation, and referral to enforcement agencies. *Currently, no information on sleep loss or fatigue is available from SENSOR programs*, but the case-based SENSOR model might be developed in the future to characterize fatigue-related cases and devise prevention programs.

Another case-based collaboration between the States and NIOSH, the Fatality Assessment and Control Evaluation (FACE) program, conducts State censuses of fatal occupational injuries and investigates selected injury types, populations, or working conditions. Some fatalities specifically investigated by FACE include: youth under 18 years of age, fire fighters, highway/street construction work zones, and machine-related injuries. Recommendations for prevention activities are based on in-depth investigations of FACE cases and series of cases of a similar type. *No information on fatigue can be obtained from current case series*, but the FACE approach might be useful in future efforts.

In summary, some specialized surveillance programs allowing follow-up or detailed case-based investigations (e.g., NEISS, SENSOR, FACE) hold promise for the development of convenient sleep loss or fatigue indicators that, with a sufficient number of cases, would allow better characterization of fatigue as a contributing factor to work-related injury or illness. Prevention programs to reduce fatigue, such as design of better work schedules, or re-organization of job tasks or workload, could then be targeted toward those occupations or worksites where surveillance has identified the highest risk. Continuing surveillance could then be used to monitor the progress of prevention programs.

In her presentation, Ms. Katharine Newman (U.S. Department of Labor Bureau of Labor Statistics [BLS]) said that the Census of Fatal Occupational Injuries (CFOI) reports comprehensive information on deaths due to on-the-job injuries. The Annual Survey of Occupational Injuries and Illnesses provides a sample of data based on OSHA 200 logs (records required by the Occupational Safety and Health Administration [OSHA]). For cases involving days away from work, the Annual Survey collects data on occupation, the number of days away from work, the nature and source of the injury or illness, the part of the body affected, and the event or exposure. For example, a truck driver may sprain (nature of disabling condition) his or her back (part of body affected) while lifting (event or exposure) a container (source directly producing disability).

Ms. Newman noted that the incidence of workplace illnesses and injuries causing lost workdays has been declining since 1990 for the labor force as a whole. Still, only a few occupations, such as construction laborers and farm workers, have higher rates of occupational fatalities than truck drivers. In 2001, truck drivers accounted for over 8% of nonfatal occupational injuries and illnesses in the U.S. but less than 3% of employment.

Highway collisions cause about two-thirds of truck driver fatalities but only about 13% of nonfatal driver injuries involving days away from work. About half of the driver injuries involving lost workdays are sprains, often caused by overexertion such as lifting heavy objects. Many truck drivers injured or killed on the job do not work in the transportation industry; rather, they drive trucks for industrial or retail companies. Furthermore, 28% of driver injuries or illnesses involving days away from work involved drivers with less than one year of tenure with their current employer. This suggests that the high turnover prevalent in the trucking industry may be an independent contributor to truck driver occupational safety and health problems.

Dr. Ralph Craft (U.S. Department of Transportation Federal Motor Carrier Safety Administration [FMCSA]) reported on truck crash data available from the Department of Transportation. The National Highway Traffic Safety Administration (NHTSA) of the U.S. Department of Transportation runs the Fatality Analysis Reporting System (FARS), which has limited information based on police reports. The University of Michigan Transportation Research Institute (UMTRI) compiles Trucks Involved in Fatal Accidents (TIFA) for the FMCSA; it supplements FARS with interviews, providing a richer data set. Both FARS and TIFA are ongoing surveillance efforts. Another rich data set is the Large Truck Crash Causation Study (LTCCS), a one-time data collection effort currently in progress. LTCCS will provide data on truck crashes causing an injury or fatality involving trucks weighing at least 10,000 pounds. Dr. Craft reported that LTCCS obtains data on the reasons for the crash, including:

- Driver errors (decision errors, performance errors, or nonperformance errors—as when a driver dies of a heart attack and then the truck hits somebody else)
- Driver fatigue
- Illegal drug use
- Alcohol use
- Prescription drug use
- Prior health history of the driver
- Driver illnesses
- Relations with motor carriers and work related pressure
- Problems with the roadway

An important empirical question is the extent to which driver fatigue causes highway crashes. Different studies yield different answers, depending on such factors as the measures of fatigue used and the nature of reporting. According to Dr. Craft, LTCCS suggests that 26% of heavy truck crashes resulting in an injury or fatality were due to environmental conditions such as ice on the roadway; 25% were due to prescription medicines, alcohol, or illegal drugs; and only 6% were due to driver fatigue. Australian data presented by Dr. Ann Williamson later in the conference, however, suggested that fatigue was involved in 49% of heavy truck crashes. A European report, based on the contributions of Drs. Torbjorn Akerstedt, Catherine Garo, Patrick Hamelin, Nick McDonald, and Freek van Ouwerkerk, said that fatigue was a significant factor in approximately 20% of commercial truck and bus crashes.[28] Without a "fatigue-alyzer" it always will be very difficult to determine the boundary beyond which fatigue becomes a safety hazard; arguably, since fatigue, sleepiness, and attentiveness are factors with continuous gradations, what limit do we want to establish for legal driving (as we do with alcohol) and how do we develop metrics and test drivers? Collecting additional high-quality data about driver fatigue could help resolve uncertainty about the causes of truck crashes.

Dr. Margaret ("Meg") Sweeney (U.S. Department of Transportation, Bureau of Transportation Statistics [BTS]) discussed data sources available from BTS. Motor Carrier Financial and Operating Statistics include data on income, tonnage, mileage, employees, and transportation equipment on Class I and Class II for-hire intercity contract and common motor carriers.* While this rich data set is available only for these relatively large carriers, the carriers are identified by name and DOT number, and data in this set can be linked to other data sets that identify the carrier. The Commodity Flow Survey reports data on five million shipments per year, including the origin, destination, and commodity shipped. The BTS Omnibus Survey is a monthly household survey with rotating questions on issues such as satisfaction with safety standards for large trucks. The BTS also has data on border crossings.

Dr. Brenda Lantz (Upper Great Plains Transportation Institute) reported on a study she conducted with Michael Blevins of FMCSA regarding commercial vehicle driver traffic convictions. The purpose of their study was to help identify motor carriers with a high risk for safety problems. They combined two data sources, drawing stratified random samples from each: driving convictions for about 65,000 commercial drivers, based on the Commercial Driver License Information System (CDLIS); and vehicle inspection and crash reports for about 14,000 carriers, based on the Motor Carrier Management Information System (MCMIS). By linking these two data sources, they found that driver conviction data serve as an indicator for carriers with safety problems. Specifically, carriers with higher rates of driver traffic convictions were also more likely to have (1) vehicles placed out of service as a result of roadside safety inspections, (2) high rates of crashes per power unit and crashes per driver, and (3) poor scores on the Motor Carrier Safety Status Measurement System (SafeStat).

Stephanie Pratt (NIOSH) compared the two primary sources of occupational fatality data on truck drivers: CFOI and FARS. The two data sets provide similar counts of the number of truck driver fatalities. According to CFOI, there were 7,228 occupational fatalities among truck drivers between 1992 and 2000, of which 4,834 (67%) were due to highway crashes. Contact with objects or equipment, assaults or other violent acts, harmful substances or environments, or falls caused about 20% of occupational fatalities for drivers.

Ms. Pratt, however, noted that CFOI and FARS differ in several respects. First, the two systems have different data structures. CFOI records are person-based, with a record for

---

* Unfortunately, this data set, which has been used widely for many decades in support of scholarship on issues ranging from motor carrier safety to productivity, was terminated in 2005.

each worker who died but no information on injured persons unless it is included in the case narrative. FARS has accident-level, vehicle-level, and person-level records, with information on other involved persons and vehicles. On the other hand, CFOI provides more detailed occupational and industry data on the decedent than does FARS, which allows researchers to identify high-risk groups of workers and develop interventions directed to these workers and their employers. CFOI also uses multiple sources to ascertain a work relationship for a fatality, whereas FARS relies primarily on death certificates. Finally, CFOI and FARS use truck classification schemes that are not directly comparable. The data structure and the availability of occupation and industry data affect the way researchers frame their analysis of the data, while the differences in truck classification and ascertainment of work relationship undoubtedly influence comparability and case counts. Ms. Pratt considers CFOI and FARS complementary data sets, both of which should be considered by researchers.

Several points related to these issues were raised by other conference participants:

- It is difficult to get data that reports the number of hours a driver has been working at the time of the crash, as well as the hours of awake time and hours of work and sleep over the previous seven days.
- Employers now use OSHA 300 logs, which provide information on the time of day of a crash and the start of the driver's work shift. It would be desirable to obtain more data, but it is difficult to get even a single additional item included in reporting forms.
- Data on work hours are not kept in machine-readable form. It would be very useful to have data not just on the work shift in which a crash occurred, but also on cumulative fatigue from the week preceding a crash.
- Dr. June Fisher noted that medical exams are required for transportation workers. The Federal Aviation Administration requires that physicians doing these exams for aviation workers be certified. These medical exams may have other relevant information.
- Dr. Ann Williamson emphasized the need to develop validated and sensitive fatigue measures.
- The lack of objective measures of driver fatigue (comparable to a breathalyzer for alcohol intoxication) makes it difficult to determine whether fatigue played a role in a crash. Fatigue is a continuum with varying effects under different circumstances.
- Dr. Michael Quinlan noted that there are no workers' compensation data for many self-employed truck drivers, who may not be covered by workers' compensation. Employed drivers who are covered also may not file claims because filing claims may endanger their future employment.
- Dr. Dieter Plehwe pointed out that fatality rates in trucking were falling in the EU.

The development of more systematic data on morbidity and life expectancy for truck drivers is an important area for future research. Several studies have shown that less privileged groups tend to die younger and have more health problems than more privileged groups. A study of French men found that managers had longer life expectancy and longer disability-free life expectancy than did manual workers.[29] A U.S. study found that age-adjusted death rates were higher for low income and poorly educated persons than for those with more income and education and that the socio-economic disparity in death rates increased between 1960 and 1986.[30] Socioeconomic differences in mortality rates in the U.S. stem from a variety of causes, including behavioral risk factors (smoking, heavy drinking, obesity, and lack of physical activity), access to medical care, exposure to occupational hazards, and stress related to having a lower position in the social hierarchy.[31,32] The Whitehall II study of British civil servants found that employees reporting a low level of control over their jobs had more sickness absences from work[33] and higher risks of developing coronary heart disease;[34] both the Whitehall II study and a study of French civil servants found that employees lower in the hierarchy were more likely to report that their health was poor.[35]

At the conference, Mr. John Siebert of OOIDA reported that owner-operator truck drivers had short life expectancies. Data on active OOIDA members who had died suggest that the average age of death for OOIDA members is only 55.7 years. While this sample is not random and is biased toward those actively in the work force and under special pressures encountered by owner-operators, at face value Mr. Siebert considered these numbers alarming. Using OOIDA member data, he reported that 49.5% of OOIDA members were obese, compared to only 31% of the overall U.S. population. Only 12.7% of OOIDA members were normal weight, compared to 36% of the overall U.S. population.

Mr. Siebert also reported that 13 out of 430 recent OOIDA deaths (3%) were from suicide.* Suicides among owner-operators may stem partly from intense financial pressures: Dr Burks said that owner-operators frequently miscalculate their costs and consume their capital (the rational choice failure discussed above). Dr. Quinlan said at the conference that there was a high suicide rate among Australian drivers. Records of the Transport Workers Union of Australia Superannuation Fund (mainly for employee drivers) for the period from July 1995 to July 1998 found that suicides accounted for 10% of recorded deaths, more than three times the Australian average. Most of the Australian drivers committing suicide, Dr. Quinlan reported, had suffered a breakdown in their marriage or relationship in the past six months and were under severe financial pressure.[23]

Mr. Siebert reported (based on a Teamster source) that union drivers had a life expectancy of 63, about seven years longer than the 55.7 for OOIDA members. These numbers merit further investigation. First, are they accurate? Second, if they are accurate,

---

* For comparative purposes, the percentage of deaths among all males in the U.S. in 2002 caused by suicide were as follows: age 25 to 34, 14.4%; age 35-44, 9.2%; age 45-54, 4.5%; and age 55-64, 1.9%. See *National Vital Statistics Reports*, Vol. 53, No. 17, March 7, 2005, Table 1, page 16, available online on July 10, 2006 at *http://www.cdc.gov/nchs/data/nvsr/nvsr53/nvsr53_17.pdf*. Using figures in this table, it can be calculated that 4.9% of all deaths in 2002 among U.S. males aged 25 to 64 were caused by suicide. The percent of all deaths caused by suicide was half as high among females as among males, but the majority of truck drivers are male.

why do Teamster drivers tend to live longer than OOIDA drivers? Are the differences in life expectancy caused by differences in working conditions and financial stress between union drivers and owner-operators? Alternatively, do pre-employment differences between union and nonunion drivers explain the apparent discrepancy between these two groups? Third, how do life expectancy figures for other drivers (those belonging to neither the Teamsters nor OOIDA) compare to those for Teamster or OOIDA members?

This discussion also suggests several possible research topics that were not explicitly addressed at the conference. Life expectancy, morbidity, and disability data specific to truck drivers but more broadly representative than data for members of OOIDA or the Transport Workers Union of Australia are sparse. It would be especially useful to have not only standardized mortality ratios for truck drivers, but also a measure similar to the one used in a recent study of uranium miners: years of potential life lost per year employed in that occupation.[36] Employment as a driver may also reduce future earnings potential by making the driver unfit for work. It would be instructive to determine if persons employed as drivers were systematically more likely than other blue-collar workers to become disabled or to retire early due to health reasons. Do drivers face a disproportionate risk of having very low incomes in their fifties and early sixties?

Another important area for future research is to replicate with American data the foreign epidemiological studies concerning the incidence of disease among truck drivers. Drs. Saltzman and Belzer summarized some of these foreign studies as follows:[37]

> Clearly, drivers disproportionately suffer from certain health problems, many with delayed onset. An epidemiological study of over 450,000 Canadian men found that truck drivers faced higher risk of death than other men did from colon cancer, laryngeal cancer, lung cancer, diabetes, ischemic heart disease, non-alcohol cirrhosis, and motor vehicle accidents.[38] A Danish study found that a group of 14,225 truck drivers had higher mortality over a ten-year period from lung cancer and multiple myeloma than did a group of 43,024 unskilled male laborers in other occupations.[39] An analysis of virtually all admissions to Danish hospitals over several years found that, compared to the male working age population, both truck and bus drivers had especially high age-standardized hospital admission ratios for lung cancer, ischemic heart disease, cerebrovascular disease, chronic obstructive pulmonary disease, and prolapsed cervical or lumbar discs; and truck but not bus drivers had especially high admission ratios for back injuries.[40]

Certainly, the obvious sources of occupational fatalities and injuries such as highway crashes and back sprains from lifting heavy objects merit the attention that they have received in data collection programs such as CFOI. However, the more subtle linkages between employment conditions for drivers and health problems such as diabetes and cardiovascular disease—linkages that are not always recognized because of long latent periods and the simultaneous presence of non-occupational causes—require more attention from future research because they take a heavy toll in morbidity and mortality.

For policy purposes, it would be very useful to combine epidemiological and economic analysis, helping researchers estimate of the dollar cost of morbidity and premature mortality associated with employment conditions for truck drivers.

Ergonomics, Job Injuries, and Exposure

Truck drivers face a variety of problems related to ergonomics, job injuries, and exposure. For example, those who drive gasoline tanker trucks often experience acute headaches, dizziness, or nausea after exposure to gasoline vapors during loading and unloading.[41] Long work hours intensify hazards of truck drivers' exposure to harmful substances or conditions, particularly since "occupational exposure limits are almost invariably calculated on the basis of an 8 hour day, 5 day week."[42]

At the conference, Drs. Eric Garshick and Thomas Smith (Harvard University) presented work they are doing with Dr. Francine Laden (also of Harvard) on the health effects of exposure to diesel exhaust. According to Dr. Garshick, 40 previous studies indicate that truck drivers and others likely to have diesel exposure have a 20% to 50% elevated risk of lung cancer. Furthermore, diesel exposure may be associated with chronic respiratory diseases such as asthma, respiratory symptoms such as wheezing, reduction in pulmonary function, and allergic inflammation. They cited a recent study reporting that residence near a roadway in the Netherlands (where diesel engines are more common than they are in the U.S.) was associated with excess mortality from cardiopulmonary causes.[43]

The current Garshick-Smith-Laden project studies over 60,000 unionized LTL trucking industry workers working for four large carriers in 1985: ABF Freight System, Consolidated Freightways (now out of business), Roadway Express, and Yellow Freight System (which bought Roadway in July 2003). Using work history data from the employers and death information from the Central States Pension Fund and the National Death Index, they are examining the relationship between mortality and the intensity and duration of exposure to sources of diesel exhaust and other sources of particulates. Based on data from a questionnaire sent to a sample of 10,000 workers, they are adjusting for differences in smoking rates by job title and work location. Their industrial hygiene sampling team visited many freight terminals to measure the exposure to diesel particles for various emission sources (engine type, fuel, maintenance), job title (long haul driver, pickup and delivery driver, dock worker, hostler,* mechanic, or other), work settings (vehicle cab, freight dock, shop, or other), terminal characteristics (size, location), and air pollution background levels. The Garshick-Smith-Laden study has several advantages over some previous studies of diesel exposure and lung cancer risk: a large cohort with computerized job history information extending over most of each worker's lifetime, and extensive sampling to provide estimates of the composition and intensity of exposure. Their project currently is funded through 2006.

Detailed work histories are important because exposure varies according to job and work location within the trucking industry. For example, diesel engines became common in long haul trucks during the 1950s and 1960s, but they did not become common in pickup

---

* A hostler (also known as a "yard jockey") often uses a specially configured tractor to move trailers in and out of freight docks and park them in the yard.

and delivery trucks until the 1970s and 1980s. Forklifts used in freight terminals used diesel fuel in the 1980s but now use propane or liquefied petroleum gas.[*] The yard hostler's entire shift is spent in a specialized hostler tractor, working around the yard, where there may be a large number of diesel tractor units starting (a smoky phase of operation) and operating. In smaller companies or terminals, hostlers may also use old road or city tractors that no longer are roadworthy, further exposing themselves to environmental hazards. He or she may also be responsible for hooking up truck-trailer combinations for road or city drivers, who then operate the truck-trailer combination on the highway. Location is important because many freight terminals are located in industrial areas with industrial air pollution, in addition to truck exhaust. Dr. Smith reported, however, that driver exposure depends more on emissions from vehicles in front of the driver's truck (which are especially intense during acceleration and deceleration) than on emissions from the driver's own truck.

The presentations by Drs. Garshick and Smith suggest some possible areas for future research that were not discussed at the conference. Can routine truck inspections readily detect diesel emission problems that can be fixed inexpensively by such measures as cleaning or replacing the injectors and sealing off leaky gaskets that leak exhaust into the truck cab? Is it cost effective to reduce diesel exhaust exposure by installing filters in the heater and air conditioner air intake systems of vehicles such as pickup and delivery trucks, urban buses, and taxis that are driven for extended periods in congested urban areas?

Dr. William Rogers (Motor Freight Carriers Association) presented his analysis of workers' compensation claims experience in unionized LTL carriers. He noted that unionized LTL carriers provide good pay (typically $60-70,000 per year for drivers), plus pensions and excellent health insurance. The union contract protects employees from management abuse. As a result, employees of unionized LTL carriers have long job tenure—often 30 to 40 year careers in one company. Dr. Rogers pointed out that many union LTL drivers have driven one million miles or more without an accident. He said it would be useful to study not only safety errors, but also the roots of success for those workers with no accidents during lengthy careers.

Dr. Rogers presented data on the incidence of workers' compensation claims. Workers with fewer than two years of job tenure account for 22% of workers' compensation claims for drivers and 24% of workers' compensation claims for other employees. Workers aged 35 or younger account for 24% of workers' compensation claims but only 15% of the workforce. Older workers probably are less likely to be injured because they know what they are doing, Dr. Rogers continued, but they take longer to recover if they are injured.

---

[*] Dr. Smith noted that Mexican freight docks generally do not have forklifts; they still load freight manually.

Dr. Rogers said that injury rates among dock workers and yard workers have been reduced by technological change. Now, most LTL freight is shrink-wrapped and palletized so that it can be handled with forklifts. This arrangement reduces the incidence of finger injuries and lifting injuries. Still, dock and yard workers are over-represented among workers' compensation claimants: they account for 35% of all claims but only 18% of the work force.

Dr. Rogers presented statistics on permanent disability claims, usually the most expensive workers' compensation cases. Approximately 10% of all 1999 claims resulted in a permanent disability, and these claims accounted for 44% of all claim costs. Payments for permanent disability claims were not evenly distributed among occupational groups, as shown below:

Table 8. Permanent Disability Claims for Unionized LTL Carriers

Occupational group	Percent of work force in unionized LTL carriers	Percent of payments for permanent disability claims for unionized LTL carriers
City drivers	41	35
Road drivers	23	32
Dock workers	15	15
Yard workers	3	11
Office workers	14	4
Mechanics	4	3

Road drivers and, especially, yard workers accounted for a disproportionate part of share of permanent disability payments. Workers aged 51 to 55 also accounted for a disproportionate share of permanent disability payments: they were 18% of the work force but accounted for 28% of payments. Workers with less than one year of tenure were another high-risk group: they had 9.2% of disability claims even though they probably comprise about 1% of the Motor Freight Carriers Association workforce.

Dr. Rogers reported that back sprains or strains, upper body sprains or strains, and knee disorders were the three most prevalent types of injuries, accounting together for about 41% of workers' compensation claims and about 44% of claims costs. He noted that average claim cost was about three times higher in cases where an attorney represents the workers' compensation claimant than in cases where the claimant is not represented by an attorney.

Dr. Rogers' findings suggest a number of potential topics for research. Are workers' compensation statistics on occupational injuries and illnesses consistent with other data sources, such as BLS's Annual Survey of Occupational Injuries and Illnesses? What changes can be made in freight yards to reduce the high rate of permanent disability among yard workers? To what extent does the higher claims cost for workers' compensation cases with attorney representation stem from attorneys selectively agreeing to represent clients who have a strong case and large potential claims? To what extent does the mere involvement of an attorney affect the outcome? Do attorneys win larger settlements than the claimants would have received without attorney representation, or are attorneys involved in cases more likely to have larger settlements?

In the discussion, Dr. Burks said that small carriers in California face very high workers' compensation premiums because of their high injury rates. They may face premiums of $33 for every $100 in payroll. According to Dr. Belzer, however, part of this additional cost may stem from unusually generous legal provisions in California that not only provide incentives for lawyers to take these cases, but also allow claimants to shop for doctors and treatments as well. Research that controls for these effects clearly is needed.

Dr. James McGlothlin (Purdue University) presented research he conducted with master's student Patrick Sheets: an ergonomic and cardiovascular stress evaluation of beverage deliverymen. They generally *are* men because one needs great shoulder strength to deliver beverages. They are paid based on how much soda and beer they deliver.

Dr. McGlothlin said that beverage delivery drivers rarely work these jobs into their fifties. Beverage delivery work can put excess strain on their cardiovascular system due to physical demands of loading and unloading, long workdays, awkward postures, and high work rates (fast-paced, intensive work). Beverage delivery drivers also face robbery hazards. Beverage delivery is mostly a cash-and-carry operation, especially for deliveries in economically distressed areas where stores do not have good credit and must pay in cash. A beer delivery driver may have $1,000 to $3,000 in cash by the end of the day, making the driver a robbery target. Beverage delivery drivers face traffic hazards as well, particularly when parked on a busy street while unloading. In winter, frozen bottles can become like hand grenades, exploding when handled and sending glass shards into the driver's face. In 2000, the number of lost workdays due to injury or illness per 100 workers was about three times as high for beverage delivery drivers as it was for all workers in private industry.

Dr. McGlothlin analyzed driver heart rates to determine which specific beverage delivery activities were most physically demanding and to determine if ergonomic redesigns could reduce cardiovascular stress. He studied nine subjects for a four-month period, recording their heart rate and videotaping them throughout the workday. On average, they handled about 36,000 pounds of beverages per day. Dr. McGlothlin said that working at more than 40% of a person's maximum heart rate for a sustained period of time leads to fatigue, and he found that unloading beverages from delivery trucks raises the driver's heart rate substantially. Increases in heart rates generally occurred during overhead work, handling of beverage containers below the waist, and pushing or pulling two-wheel hand trucks up inclines or stairs.

Dr. McGlothlin considered several changes that could reduce strain on beverage delivery drivers. One is to use plastic bottles instead of glass, as plastic weighs less. He noted, however, that beverage companies prefer to force plastic bottle suppliers to compete with glass bottle suppliers in order to keep bottle prices down. Ergonomic changes to the delivery truck could also be made, such as:

- Pull-out step-on platforms for unloading beverages (which reduce the distance that the driver must reach to obtain the beverage containers and allow the driver to place the containers on the platform for subsequent placement onto the hand truck)
- Proper tire pressure on hand trucks (facilitated by placing an air compressor next to the end of the row where drivers obtain hand trucks)
- Exterior grab handles by all bays
- Three-position drop shelf holes for all bays
- A wider step platform on the housing step bar
- New rollers in all bay door slats, and lubricated doors
- Air-cushioned driver's seat

He found that, for six of the nine subjects, these ergonomic interventions reduced the percent of the time that their heart rates exceeded 50% of their maximum heart rate; two of the remaining three subjects never exceeded 50% either before or after the interventions. A potential future research project might replicate the McGlothlin study of beverage delivery drivers with other groups of truck drivers, particularly in LTL and package delivery.

Dr. Michael McCann (Center to Protect Workers' Rights) gave a presentation on occupational hazards of ready-mixed concrete truck drivers. A more detailed report on this subject, submitted to Dr. McCann by Nancy Clark, Jonathan Dropkin, and Lee Kaplan of the Mount Sinai-Irving J. Selikoff Center for Occupational and Environmental Medicine in New York City, is included on the accompanying CD-ROM and at *http://www.ilir.umich.edu/TIBP/truckdriverOSH*. The Clark *et al.* report stated that ready-mixed concrete truck drivers have several job duties, including loading and mixing concrete at the plant, delivering the concrete to the construction site before it sets, and cleaning the mixer drum. The industry had a higher rate of OSHA-recordable nonfatal occupational injuries and illnesses than construction as a whole. Clark *et al.* analyzed OSHA 200 log forms for 23 ready-mixed concrete plants between 1997 and 1999, finding that sprains and strains accounted for 62% of lost-time cases. Other problems included bruises, eye injuries from particles, cuts, hearing loss, fractures, and amputations.

Based on their site visits and their review of previous research, Clark *et al.* listed several occupational hazards faced by ready-mixed concrete truck drivers. Slips, trips, and falls accounted for about half of the injuries, according to a 1986 study. They wrote that "hazards include slippery surfaces, unsure footing, damaged ladders and walkways, and unsure hand- and footholds during climbing and descending truck cab and equipment."[44] Drivers can be struck by or pinched by equipment or materials. They face ergonomic risk factors such as:

whole body vibration from driving the trucks, awkward and fixed postures (for instance, while hosing down the inside of the truck and holding the driving wheel of an empty truck over bumpy roads), forceful muscular activities (for instance, lifting heavy chutes, frequently lifting chutes, and activating a drum when discharging concrete), extremes in temperature (hot and cold), and repetitive twisting of the back and neck (for instance, when delivering concrete or looking out the back of a truck).[45]

Ready-mix operations are noisy. Drivers' noise exposure exceeded the NIOSH recommendation of no more than 85 dB averaged over eight hours. When cleaning the drum every few months, the drivers or contract workers must work within a confined space in which they are exposed to noise and silica dust. During loading and unloading, they may be exposed to irritating chemicals. They may be burned by hot surfaces, face eye injuries from flying particles, and face rollover hazards "while driving or unloading on unstable, uneven or steep ground at delivery sites."[46]

Clark *et al.* recommended a number of measures to prevent occupational injuries and illnesses among ready-mixed concrete truck drivers. These included safety and health training for workers and managers, active participation of workers and unions in identifying and controlling hazards, redesigning equipment to reduce falling hazards, having machine guards and lockout/tagout, better seats to reduce exposure to vibration, and special confined-space programs for cleaning out mixer drums.

Some of the hazards that Clark *et al.* identified for ready-mixed concrete truck drivers also affect other drivers and merit further research. Many drivers face extended exposure to highway noise, which can lead to hearing loss,[47] a problem exacerbated when drivers sleep in their trucks while their partners drive and thus lack recovery time between exposures.[48] Can changes in cab and sleeper berth design reduce noise exposure? Many drivers are exposed to whole body vibration,[49] which can lead to low back pain.[50,51,52] Can changes in seat or sleeper berth design reduce exposure to vibration?

A number of other exposure hazards were raised by conference participants. Dr. Michael Quinlan said that there was a need for more research on occupational violence in trucking. In 2001, Dr. Quinlan and his colleague Claire Mayhew surveyed 300 Australian drivers about this and other topics. About three in ten drivers reported being subjected to verbal abuse, somewhat fewer than one in ten reported being threatened, and 1% reported being assaulted. At freight forwarding yards, verbal abuse and threats were closely linked with economic pressures in nearly all incidents. Loading delays, drivers cutting in line, and mistakes by forklift drivers fueled tensions, which led to the violent behaviors. Other occupational safety and health problems that drivers reported in the Mayhew-Quinlan survey include crashes (14% in the past year and 24% in the past five years), acute injuries (over 25% in the past year, mostly minor), chronic injuries (over 50%, mostly back), and hearing loss (22%). Dr. Quinlan's presentation at the conference also reviewed the literature on driver occupational safety and health problems, noting studies of vibration, back, and spinal injuries; hazardous substance exposure; and crashes.

Dr. Stephen Burks commented on the higher injury rate for less experienced workers. Many drivers quit after only a few months, so that trucking has many inexperienced workers. He called for more research on how to reduce driver turnover, as reduced turnover would improve safety. He said that there were pockets of high injury rates in trucking, such as in plate glass transport that merited further study.

These comments raised additional important subjects for future research. Can more intensive training and supervision reduce the injury rate among inexperienced drivers, dockworkers, and yard workers? How can work be redesigned to reduce the negative health and safety consequences of current work methods? How can trucking companies establish a "corporate culture" that values safety?

Labor Market, Employment Relations, and Personnel Management

Dr. Daniel Rodriguez (University of North Carolina-Chapel Hill) presented the results of a study he conducted with Dr. Michael H. Belzer (then of the University of Michigan and now at Wayne State University) and Dr. Stanley A. Sedo (University of Michigan) concerning the relationship between driver pay and highway safety. Traditional approaches to safety emphasize human factors, load characteristics, vehicle characteristics and maintenance, and environmental conditions. Belzer, Rodriguez, and Sedo considered whether safety depends on economic factors such as form of driver pay (hourly pay rather than mileage or percentage of revenue, both of which are "piecework"), the level of driver pay, and the impact of product market competition in trucking on the human capital of truck drivers.

They did three empirical studies. First, they analyzed data for the period from 1995 to 1998 on more than 11,000 drivers working for J.B. Hunt, the second largest truckload carrier in the U.S. Hunt raised driver wages an average of 38% early in 1997 in an effort to reduce high turnover and improve safety and reliability. Second, they studied data for a cross-section of 102 truckload carriers in 1998. Third, they used data from a 1997-98 survey that included 247 mileage-paid employee drivers working in the for-hire trucking industry.

In all three studies, they found that higher truck driver pay was associated with a lower rate of crashes. In the J.B. Hunt study they found that, starting from the mean mileage rate of 34 cents per mile, a 10% higher base mileage rate was associated with a 34% lower probability of a crash. In addition, a 10% increase in driver pay was associated with a 6% lower crash probability, giving an overall pay-rate effect of 1:4. In the cross-sectional study of 102 truckload carriers, they found that – at the mean – for every 10% more in average total compensation for drivers, the carrier would experience a 9.2% lower crash rate. In the driver survey study, they found that drivers earning a 10% higher mileage rate would have an 18.7% lower probability of having a crash during the reporting year, while drivers having 10% more paid days off would have a 6.3% lower probability of having had a crash during that same year. They gave two possible explanations for this effect of pay on driver performance. First, higher compensation allows carriers to recruit and retain better-qualified drivers (a human capital effect). Second, higher compensation may have provided an efficiency wage incentive for drivers to drive safely, lest they lose a job that pays better than their alternative wage.

Dr. Rodriguez and his colleagues also examined the impact of pay rates on the number of hours that drivers wished to work. Based on the driver survey data, co-author Stanley Sedo found a classic backward-bending labor supply curve. When the pay rate was very low, an increase in the pay rate made drivers want to work longer hours because the pay increase raised the implicit price of leisure (the substitution effect). When the pay rate was high, however, an increase in the pay rate made drivers want to work shorter hours because they could now afford to buy more of all normal goods, including leisure (the income effect). The income effect becomes dominant—i.e., higher pay begins to reduce rather than increase the number of hours that drivers wish to work—when the mileage rate paid to drivers increases above 31.4¢ per mile. In order to align the drivers' economic preferences with public safety policy preferences, and reduce the number of hours drivers want to work per week to 60 (the legal maximum work hours for truck drivers since 1938), the mileage rate might be raised to 37.8¢ per mile. Their driver survey, however, showed that the average mileage rate for drivers with three years experience was only 28.6¢ per mile, which helps explain why many drivers work more than the legal maximum set by HOS regulations.[21, 53, 54]

While Belzer, Rodriguez, and Sedo demonstrated that truck driver compensation is a powerful predictor of both driver and motor carrier safety, limitations in the data prevented them from determining whether the value of this safety improvement exceeded the costs. The Trucking Industry Benchmarking Program, developed and directed by Dr. Belzer, was designed to link driver characteristics and motor carrier operations to safety performance, allowing a much more accurate weighing of factors operating simultaneously in the industrial environment. With information that allows researchers to weigh relative effects of various carrier-level safety measures, the predictors of safety performance will be much better understood (See *http://www.ilir.umich.edu/TIBP*). Further research is needed, however, to show the value of safety return for every dollar spent on increased driver wages. Does safety pay? If it does not, will consumers and government policy-makers be comfortable paying for the marginal cost of an additional safety margin?

Dr. Ian Savage (Northwestern University) presented his research on the economics of transport safety. He compared FMCSA to a policing organization because of its emphasis on activities such as roadside inspections to catch unsafe trucks. He argued that safety fears expressed at the time of trucking deregulation were overblown. Although there had been fears of "killer trucks" at the time of deregulation, the rate of fatal truck crashes per mile driven has actually fallen since then. Nevertheless, safety advocates – and particularly families of victims – call for further improvements in truck safety.

Dr. Savage asserted that there is no universal answer to whether shippers prefer high safety at a high price, or low safety at a low price. Safety is expensive to provide. Although shippers prefer more safety to less, their willingness to pay depends on the reduction in expected harm. Electronics shippers have a delicate product for which shipping costs are a small fraction of the final price; they therefore want safety and will pay for it. Gravel haulers, in contrast, have a product that is not damaged if the truck is in a collision, and shipping costs are a substantial fraction of the final price. They care more about low shipping cost than they do about safety.

In order to justify government safety regulation, Dr. Savage continued, one must identify a market failure. Imperfect information is a problem: insurance carriers need to tailor rates to the risk of the trucking company, but they often say it is too costly to assess the risk of individual carriers. Similarly, if customers for trucking services are poorly informed, then unsafe carriers may be able to charge high prices, pretending to be safe firms. Dr. Savage therefore argued it is appropriate for government to inform shippers about the safety records of different carriers. He also said that it is appropriate to require a minimum level of safety because of externalities. He noted that the government does this by setting minimum standards for safety *inputs*, using these as an approximation for the minimum acceptable safety *output* level. But if shippers have different tastes for safety, Dr, Savage concluded, it is appropriate to have variation among carriers in safety levels—and the prices that they charge for their services.

Combining Dr. Savage's point about differing customer preferences regarding the tradeoff between safety and shipping cost and Belzer *et al.*'s finding that higher wages at J.B. Hunt reduced crash rates, it would be instructive to examine changes in J.B. Hunt's customer mix after they raised wages. Was Hunt able to attract safety-conscious shippers, for whom high standards of transportation quality are important? Indeed, that was an important part of their strategic decision to become a quality operator. In addition, since the damages produced by the crash of a gravel-hauler may equal or exceed those produced by an electronics hauler (to use Dr. Savage's example), how do markets compensate those who are harmed by the crash? In other words, if the gravel-hauler has less insurance coverage than the electronics hauler (consistent with the lower price charged to haul the goods), will the public bear a larger fraction of the cost of a gravel-hauler's crash than that of an electronics hauler? Will the gravel customer's lower taste for safety become a market externality – a cost borne by the public?

Dr. Belzer noted in the discussion that the LTL sector has the best safety record in trucking, and produce trucks (whose drivers have among the lowest pay rates) have the worst. Dr. Burks added that, when prices for trucking services were regulated, the Interstate Commerce Commission used to require that there be a detention charge, which provided additional compensation to carriers in the event of undue delays at the location of the shipper or consignee or delays in when the truck was permitted to enter the shipper's or consignee's premises. This requirement, Dr. Burks said, was bargained away after deregulation, and shipping contracts now typically do not include detention clauses. Drs. Belzer and Quinlan argued that restoring detention charges would reduce safety risks, particularly if drivers were paid for this time and therefore logged it as on-duty time.

Dr. Michael Quinlan's second conference presentation addressed more broadly the occupational safety and health issues prevalent in developed countries generally and Europe and Australia specifically. He briefly mentioned additional factors that affect truck safety, including:

⇒ Time pressure from elaborate supply chains and just-in-time inventory systems
⇒ Truck size, design, maintenance, and performance
⇒ Driver characteristics, selection, training, and behavior
⇒ Highway and bridge design and maintenance
⇒ Traffic levels
⇒ Behavior of other road users
⇒ Ineffective regulatory regimes

All of these contribute to health and safety concerns for truck drivers in these countries. Despite the progress made for other occupations, truck drivers lag behind in indices of health and safety. He placed special emphasis on economic forces compelling fierce competition in trucking – forces responsible for inferior health and safety outcomes for truck drivers in these countries.

Dr. Patrick Hamelin (National Research Institute for Transport Safety in France [INRETS]) gave a wide-ranging presentation on trucking in Europe. He said that long-distance truck drivers experience extreme working conditions, almost like running a marathon. As Dr. Hamelin noted in a newsletter article[55] (included in the accompanying CD-ROM and at *http://www.ilir.umich.edu/TIBP/truckdriverOSH*), drivers frequently face obstacles to staying on schedule: traffic jams, bad weather conditions, customs delays at borders, and queues for loading bays or forklifts. Often, Dr. Hamelin wrote, drivers make up for these delays by working during scheduled rest periods. Improved communications, meanwhile, have allowed employers to track the locations of their trucks, resulting in a decrease in driver autonomy.

Dr. Hamelin conducted directly comparable surveys of French truck drivers in 1983, 1993, and 1999. Drivers, he reported, work longer than average hours and earn less than other blue-collar workers do, per hour worked. Dr. Hamelin presented the following data on weekly work hours for French truck drivers in 1999:

Table 9. French Truck Driver Work Hours

Employee Group	Average Hours Worked Per Week
Drivers returning home each night and:	
Employed by private carriers	43.2
Employed for-hire carriers	47.2
Owner-operators	52.9
Drivers away from home 1-3 nights per week and employed by for-hire carriers	53.8
Drivers away from home 4+ nights per week and employed by for-hire carriers	56.1

Between 1983 and 1993, average work hours declined for drivers employed by private carriers, though this was partly offset by a shift in truck driver employment from private carriers (where hours were shortest) to for-hire carriers. Between 1993 and 1999, the number of French drivers getting sleep breaks of at least 9 hours increased while the average weekly work hours for employee drivers decreased. Work hours increased between 1993 and 1999, however, for owner-operators who returned home each night. Another development was an increase in "slip seating" (having multiple drivers use the same truck to increase the intensity of capital utilization and to keep the truck moving around the clock, which shortens delivery cycles).

Dr. Hamelin noted that tensions related to driver workloads led French truck drivers to strike in 1992. Transport prices fell following deregulation, leading to pressure on drivers to increase their workload. The pressure on drivers became a double bind in 1992, when a new demerit point system increased the risk that drivers could lose their commercial driver's licenses for driving excessive distances, driving at excessive speeds, or overloading. French truck drivers reacted with a spontaneous strike, stopping their trucks in the middle of roads and snarling traffic for 12 days. As a result of the ensuing negotiations, the 1992 regulations were revised to emphasize instead better training to qualify for a commercial driver's license.

Dr. Hamelin reported that European Economic Community (EEC) Regulation 3820/85 (adopted in 1985) limits truck drivers to an average of 45 and a maximum of 56 hours of *driving* per week. He noted, however, that this regulation allows them to have substantially longer *total* work hours because about 30% of long-distance driver time is spent in activities such as loading, unloading, waiting, vehicle maintenance, and running to find forwarding agents (a figure very similar to that found by the University of Michigan Trucking Industry Program's Truck Driver Survey, and by Martin Labbe's survey on behalf of the Truckload Carriers Association). The "average" rule allows them to work a total of about 65 hours per week, while the "maximum" rule allows them to work around 80 hours per week.

A new EU directive limits total work time for drivers to 48 hours per week. If, however, they continue driving an average of 45 hours per week, then Dr. Hamelin's data show that their total work hours will almost surely exceed 48—unless there are very big changes in transport organization and working conditions. Even if the proportion of long-distance driver work time spent on non-driving activities were reduced from 30% to 20%, 45 hours per week of driving would imply 56 hours per week of total work. Dr. Hamelin thinks it would be difficult to reduce non-driving work to less than 20%, on average, of total work time for long-distance drivers.

The differences between EU and U.S. rules for truck driver hours might make it useful to compare European and American accident and occupational illness rates for truck drivers, as well as life expectancies. Illness and life expectancy reflect the cumulative effects of many years of experience, including prior years in which work hours for European truck drivers were not as different from those of American drivers as they are today. Accident rates, however, may respond quickly to changes in work hours, so that comparing the "fixed effect" change in accident rates in the EU before and after the new EU directive on driver work hours to the change in accident rates in the U.S. during the same time span might provide an estimate of the impact of driver hour regulations on trucking safety.

Dr. Gerald Krueger (Wexford Group International) prepared a conference presentation (which he was unable to deliver in person) concerning driver wellness training programs that he conducts for truck drivers and trucking company managers. Dr. Krueger also prepared a handout (included on the accompanying CD-ROM and at *http://www.ilir.umich.edu/TIBP/truckdriverOSH*) about a driver alertness-training program sponsored by FMCSA and the American Trucking Associations (ATA). Since

1996, Dr. Krueger has repeatedly conducted a train-the-trainer driver alertness course for trucking industry trainers, safety managers, and drivers. The course notes the hazards of driving while fatigued, presents relevant findings from sleep research, and identifies what he argues are proven fatigue countermeasures.

In the discussion, Dr. Burks noted the differences among driver training providers. There were cutbacks in federal government support for commercial driving training schools after scandals arose involving some for-profit training schools. Dr. Burks argued that third-party for-profit driver training schools might not have a strong incentive to do a good job. He suggested that it may be better to have commercial driver training conducted by large trucking firms (e.g., Schneider National) that internalize costs, or else have a six-month certificate program for commercial drivers at community colleges. Dr. Belzer commented that it costs $3,000 to $7,000 per driver to hire a truckload driver. The 100% annual turnover typical in the TL sector thus is an economic problem.

Dr. Rogers said that, subsequent to the period that Drs. Belzer, Rodriguez, and Sedo studied, J.B. Hunt cut the mileage rate that they had raised in 1997. The low unemployment rate during the late 1990s and 2000 made it difficult for trucking companies to recruit and retain qualified labor unless they raised wages. The higher unemployment rate since 2000, however, reduced the pressure on carriers to raise wages and has reduced wages at some carriers.

The U.S. market for truck driver labor may soon be affected by an important development not addressed in the conference presentations: an influx of Mexican trucking firms into the U.S. market for trucking services. Cabotage rules first established by the U.S. Immigration and Naturalization Service and now enforced by U.S. Immigration and Customs Enforcement limit Canadian and Mexican drivers and carriers to shipments whose origin or destination is in Canada or Mexico; they may not handle shipments between two points in the U.S. While Canadian trucks have been allowed to travel throughout the U.S. on this basis, the same privilege has not been accorded Mexican trucks and drivers. In February 2001, an arbitration panel established under the provisions of the North American Free Trade Agreement (NAFTA) ordered the U.S. to allow Mexican trucks to operate throughout the U.S., and not just in a twenty-mile strip along the border. The U.S. DOT prepared to implement the policy ordered by the arbitrators, rather than exercise the option provided by NAFTA to keep Mexican trucks out and provide Mexico with compensation.[56] The opening of the border was delayed by a federal Circuit Court ruling that DOT had failed to conduct a legally required environmental analysis,[57] but the U.S. Supreme Court reversed this ruling in June 2004.[58] The opening was delayed further by a January 2005 report by the DOT Inspector General recommending that U.S. officials conduct safety inspections of Mexican trucks before they enter the U.S.[59] Still, it seems likely that Mexican drivers soon will be able to handle many shipments now handled by U.S. drivers. Competition from the large potential pool of low-wage Mexican drivers working throughout the United States will tend to drive down wages and other forms of compensation in the U.S. labor market for truck drivers. If the research by Belzer *et al.* is true, this may be accompanied by a substantial increase in truck crashes and other incidents.

Dr. Hamelin noted that international economic integration could have similar effects in Europe. Trucking costs are substantially lower for Hungarian and Polish firms than for Western European firms because of lower pay and social security costs.

In summary, it is appropriate to take a systems approach, considering the simultaneous use of several approaches: training drivers how to drive safely and manage their fatigue, having enough truck stops along the highway, having reasonable schedules and pay systems, and having equity in pay between U.S. and Mexican drivers. The most systematic approach to safety and health might involve making an intense effort to incorporate into the market for trucking services all relevant costs and benefits, including those that now are often ignored by decision makers because they entail externalities affecting third parties.[*]

---

[*] The Transportation Research Board Committee on Trucking Industry Research has been working with other committees for many years to get these issues on policy makers' "radar screens." See presentations from several Annual Meetings at
http://www.ugpti.org/trb/.

Fatigue and Truck Driver Safety and Health

A substantial scholarly literature demonstrates the adverse impact of fatigue on job performance. A review by Folkard and Tucker reported that the risk of accidents and injuries was approximately 30% higher for night shift workers than for day shift workers.[60] Productivity in manufacturing, they reported, was about 5% lower at night.[*] They also found that the risk of accidents and injuries during the 12th hour of a work shift was more than double the risk during the first 8 hours. Tucker, Folkard, and McDonald examined the impact of rest breaks on risk.[61] They analyzed three years of records for a car assembly plant in which workers had a fifteen-minute break after each two hours of continuous work. The risk of accidents and injuries was more than twice as high for the period from 90 to 120 minutes since the last break as it was for the period from 0 to 30 minutes since the last break. Folkard and Lombardi argued that models of fatigue and performance must reflect several risk factors for accidents and injuries: night work, an increase in the number of successive night shifts, an increase in the number of hours on duty, and an increase in the number of minutes since the last work break. They proposed a risk index adjusted for these factors as a way of assessing the risk of different work schedules.[62]

Some studies have addressed the issue of recovery from physical and mental fatigue and how this affects performance. Dinges et al. found that restriction of sleep to four or five hours per night causes significant deterioration in reaction time performance within three nights; and after a week of partial sleep deprivation, it takes two full nights of sleep for reaction time performance to recover.[63] Such partial sleep deprivation is a regular fact of life for most truck drivers. Folkard and Akerstedt, analyzing the impact of fatigue on accident risk, stated "sleeps that occur at abnormal times of day may be less restorative than those at night, even if the sleep duration is the same."[64]

The literature shows that the adverse impact of fatigue on performance applies to commercial drivers, too; see Freund for a review of the literature related to HOS regulation, including the impact of fatigue on driver performance.[65] Hamelin found that accident risk for truck drivers depended on both time of day and work span duration. For drivers with work shifts of less than 11 hours, the accident risk rate was 175% higher from 8 PM to 7 AM than from 8 AM to 7 PM (1.85 vs. 0.74); and for drivers with work shifts of 11 hours or more, the accident risk rate was 76% higher from 8 PM to 7 AM than from 8 AM to 7 PM (2.37 vs. 1.35).[66] A consensus statement by Akerstedt, Cziesler,

---

[*] In factory settings, part of the difference between day shift and night shift outcomes may stem from differences in worker experience rather than differences between day and night. High seniority workers usually get first preference for shifts and may prefer the day shift, leaving the night shift populated by less senior, less experienced workers who may be more prone to mishaps. In trucking, however, much of the over-the-road hauling in LTL (historically some of the best-paying work) is night work, and drivers with seniority tend to bid the road at least as often as the city. The seniority effect thus may bias estimates of the impact of night work per se on mishaps, but the bias is likely to be much smaller in studies based on trucking data than those based on factory data.

Dinges, and Horne (endorsed by 17 other experts) addressed accidents and sleepiness.[67] They stated that (1) fatal sleep-related highway crashes are especially frequent between midnight and 6 AM, (2) official statistics may understate the importance of fatigue in transportation accidents, and (3) countermeasures are needed to prevent fatigue from causing accidents. Stoohs *et al.* reported that sleep-disordered breathing among long-haul truck drivers doubles their accident rate.[68] Folkard stated that circadian rhythms, rather than different driving conditions at night, were probably the cause of the peak in road transport accidents at 3 AM.[69] Arnold *et al.* surveyed Australian truck drivers, finding that those who reported sleeping 6 hours or fewer prior to their current trip were significantly more likely to report hazardous events related to fatigue during that trip, such as nodding off while driving.[70] McCartt *et al.* surveyed U.S. long-distance truck drivers; 25% reported falling asleep at the wheel in the past year, and 47% reported having fallen asleep at the wheel sometime during their driving career.[71] Drivers especially likely to report falling asleep at the wheel included those with demanding work schedules (driving more than 10 consecutive hours per day, driving many hours per week, and taking fewer than 8 hours per day off duty), and those with split off-duty periods (and hence, fewer hours in the longest sleep period). Consistent with McCartt's findings, Dr. Hamelin reported at the conference that over half of the truck drivers he has surveyed have dropped off at the wheel at some time during their careers.

Numerous other studies found that driver fatigue and reduced driver alertness cause highway collisions and near misses. Among them are articles by Lyznicki *et al.*,[72] Dement,[73] Barach *et al.*,[74] Summala and Mikkola,[75] Hakkanen and Summala,[76] Philip *et al.*,[77] Feyer,[78] Horne and Reyner,[79] and Horne and Reyner.[80]

Dr. Kathryn Reid of Northwestern University made a conference presentation comparing the effects of fatigue and alcohol on highway safety, extending the work presented in the paper she co-authored with Drew Dawson.[81] Dr. Reid argues that we need an easily grasped index of impairment. Comparing fatigue to alcohol intoxication, Dr. Reid argued, makes the consequences of fatigue clear to the public and to policy makers. Impairment due to fatigue can be assessed with neurobehavioral lab tests (OSPAT or PVT); driving simulators for cars, trucks, or trains; or measures of actual job performance (e.g., observed performance driving a truck, number of accidents, or productivity). If experimental subjects have been awake for 23 hours, then their performance on psychomotor skills deteriorates to the equivalent of 0.10% blood alcohol concentration (BAC).[*]

Dr. Reid noted that the amount of deterioration in performance caused by fatigue depends on the nature of the task. Being awake 23 hours causes more deterioration than 0.10% BAC does on grammatical reasoning speed, but less than 0.10% BAC does on

---

[*] Care should be taken whenever making the comparison between alcohol and fatigue because they have different sources and progressions. Over time, intoxication subsides, but fatigue increases.

grammatical reading accuracy. Dr. Reid said we need more research about which tasks are most impaired by fatigue. She said that job performance deteriorates greatly in the first three days working on a night shift. After that, people adapt; but they still do not perform as well as on the day shift.[82]

Based on both her own research and on other studies that she reviewed, Reid concluded in her presentation (see the accompanying CD-ROM or *http://www.ilir.umich.edu/TIBP/truckdriverOSH*) that "performance is significantly impaired with moderate levels of fatigue (17-24 hours of wakefulness), equivalent to the performance seen at legally prescribed blood alcohol levels (0.05 – 0.1% BAC). The type of task and how fatigue is accrued are important in influencing the degree of impairment."[83, 84, 85, 86, 87, 88, 89] "Combining alcohol and fatigue increases the level of impairment."

Dr. Reid thought that many years of research would be needed to develop a good measure of fatigue. There is no simple marker of fatigue comparable to the breathalyzer test for alcohol intoxication. Laboratory studies can measure the impact of fatigue on various performance tasks, but laboratory performance may differ from on-the-job performance as a truck driver. It is therefore important to conduct both simulator studies and studies of actual job tasks. Also, it is difficult to determine the exact effect of a change in reaction time on the probability of a highway crash. The consequences of fatigue depend on the circumstances; falling asleep while driving is far more dangerous than falling asleep waiting at the loading dock while stationary. Finally, she saw a need for further studies of recovery from sleep loss, particularly in cases where sleep duration varies from night to night.

Dr. Hans van Dongen (University of Pennsylvania at the time of the conference; now at Washington State University) presented work he completed with colleagues regarding chronic sleep loss and wake state instability. His summary (included on the accompanying CD-ROM and at *http://www.ilir.umich.edu/TIBP/truckdriverOSH*) was as follows:

> Chronic sleep loss is common in today's 24/7 society, of which transportation operations are an integral part. Recent laboratory experiments have demonstrated that chronic sleep loss can lead to cumulative deficits in key aspects of neurobehavioral performance: sustained attention, working memory, and cognitive throughput.[90,91] These deficits are substantial: after 14 days with sleep restricted to 4 hours per day, average impairment levels are equivalent to those observed after two nights of total sleep deprivation.[90] This indicates that people do not easily adapt to the adverse effects of chronic sleep restriction. Self-reported sleepiness does not appear to track the cumulative performance deficits resulting from chronic sleep loss, however. While research subjects in a partial sleep deprivation study reported feeling somewhat sleepier during the first days of sleep restriction, their sleepiness levels did not further increase across subsequent days.[90] This suggests that people may feel they

have adapted to chronic sleep loss even when their performance deficits continue to increase, resulting in misperception of performance capability.

The discrepancy between objectively measured deficits from chronic sleep loss and subjective experiences of sleepiness might be explained by the nature of the performance impairment resulting from sleep loss. This is not just a gradual reduction of the various aspects of neurobehavioral function (e.g., cognitive slowing, diminished accuracy), but rather a rapidly fluctuating mixture of normal performance with acute lapses in performance. This "wake state instability"[92] causes people with moderate sleep loss to perform optimally *most* of the time—but not always and not reliably. Thus, while sleep-deprived individuals may not experience any immediate impairment of neurobehavioral function, critical lapses may occur in circumstances requiring sustained cognitive accuracy and speed.

The phenomenon of wake state instability, or "within-subjects variability" of neurobehavioral performance under conditions of sleep loss, makes it difficult to evaluate, in advance, whether a person can maintain optimal performance capability over time. Moreover, people differ substantially in the amount of sleep they need on a regular basis to maintain normal performance capability,[93,94] as well as in the severity of performance impairment they exhibit when getting insufficient sleep.[94] This "between-subjects variability" makes it difficult to assess whether a specific individual has obtained enough sleep to perform optimally. Little is known about these two sources of variability in the detrimental effects of chronic sleep loss on neurobehavioral functioning. Yet, understanding and predicting within-subjects and between-subjects variance is essential with regard to safety in transportation operations. Therefore, identifying behavioral or biological markers of wake state instability and inter-individual variability in vulnerability to impairment from chronic sleep loss should be key components of future research on truck driver fatigue and safety.

Advance warning of wake state instability, Dr. van Dongen concluded, would allow truck drivers to stop for a nap before serious performance lapses occur.

Dr. Orfeu Buxton (University of Chicago at the time of the conference and now at Harvard Medical School) presented evidence that endocrine function was closely related to sleep and sleep quality.[95] He defined "sleep debt" as semi-chronic, partial sleep restriction and reported that sleep debt may lead to an increased risk of obesity and diabetes. He cited the finding by Spiegel *et al.* that sleep debt has a broad impact on metabolic and endocrine systems, inducing a pre-diabetic state.[96] Chronic partial sleep deprivation was associated in this study and others with impaired glucose metabolism, abnormal cortisol (stress hormone) regulation, altered growth hormone (GH) profiles,[97] altered autonomic function (elevated sympathovagal balance), and impaired immune function.[98]

Dr. Buxton noted that poor sleep quality (which entails a loss of slow-wave sleep, or SWS) and aging can reduce GH secretion, leading in turn to loss of leanness. He speculated that sleep loss could have harmful metabolic and endocrine effects even if it does not lead to subjective sleepiness, as some sleep deprivation experiments have been able to dissociate objective and subjective sleepiness, and subjective sleepiness reports are often unreliable. He also noted that sleep loss might play a role in a metabolic syndrome of obesity, insulin resistance, and hypertension that is becoming an increasingly common and intractable health problem.[*]

Dr. Buxton outlined a number of needs for future research. One is for studies of the relationship between sleep debt and both obesity and metabolic function among workers in transportation, health care, and other industries where extended work and habitually restricted sleep are common. Second, there is a need for experiments to determine how many hours in bed a person needs on the weekend to recover the sleep loss from the preceding work week. Third, longitudinal studies of sleep and health are needed. Finally, Dr. Buxton noted a need for mechanistic studies of how sleep loss impairs metabolism and may promote obesity, insulin resistance, and hypertension so that rational counter-measures can be developed.

Studies by Stoohs and colleagues reported that sleep-disordered breathing is a common problem among long-haul truck drivers[99] and that this problem increases significantly after six days of sleeping only four hours per day.[100] At the conference, Dr. Christopher Drake (Henry Ford Hospital) presented evidence on the causes and consequences of obstructive sleep apnea (OSA), and the implications of OSA for commercial truck drivers. In OSA, a sleeping person's airway closes, causing cessation of breathing for at least 10 seconds and sometimes more than a minute. Obstructive sleep apnea, Dr. Drake noted, results in functional sleep deprivation, in which a person spending 8 hours in bed might get the equivalent of only 5½ hours of sleep. Airways are more likely to collapse during rapid eye movement (REM) sleep, when a person loses muscle tone. This is particularly important because OSA is positively correlated with obesity and large neck size, two indicators that are prevalent among truck drivers. The presence of OSA also is associated with daytime sleepiness (or even worse for truck drivers who work all night: nighttime sleepiness that is compounded by the attempt to defy the circadian sleep cycle.)

Dr. Drake commented that there is a low correlation between subjective sleepiness and objective measures of performance decrements due to sleep debt. Drivers may not know that it is essential to take a nap. An article by Dingus, Hardee, and Wierwille supports this claim, noting that "a relatively large percentage of drivers are unaware of the onset of drowsiness while driving."[101] Dr. Drake added that some individuals are less skilled than

---

[*] A scholar not at the conference, Dr. Steven Lockley of Harvard University, suggested that diabetes for shift workers may be due to eating meals at the wrong circadian phase, rather than to sleep disruption per se. See Steven W. Lockley, "Functional Physiology of the Circadian Clock," presentation at National Science Foundation Chautauqua course, "Circadian Biology: From Clock Genes and Cellular Rhythms to Sleep Regulation," May 30, 2003, Harvard University.

others in detecting that they are sleepy. Drivers who fail to recognize the symptoms of fatigue, Dr. Drake said, were the ones most likely to have accidents. He stated that a high percentage of truck driver fatalities were associated with drivers falling asleep at the wheel. In support of this concern, Dr. Torbjorn Akerstedt commented that young workers and male workers were more likely to have accidents on the night shift.[*]

It would be useful to obtain data on actual sleep for different subgroups of drivers—e.g., union vs. nonunion, employee drivers vs. independent owner-operators vs. dependent owner-operators who regularly work for the same shipper, younger drivers vs. older drivers, and drivers who work in various sub-sectors of for-hire and private trucking. Subgroup analysis might identify drivers particularly in need of fatigue countermeasures. It is known, for example, that older people have greater difficulty than younger people have in sleeping at an adverse time in the circadian cycle. On the other hand, there is a self-selection effect: those who continue to work as truck drivers are systematically more likely than newly hired drivers to be able to tolerate typical driver work schedules because those with work schedule tolerance problems eventually find other jobs.

Dr. Torbjorn Akerstedt (Karolinska Institute, Sweden) made a conference presentation reviewing the literature concerning shift work effects on health and wellbeing.[†] Dr. Akerstedt said that there have been few studies on shift work and mortality subsequent to the 1972 paper by Taylor and Pocock, which found no effect.[102] Still, Dr. Akerstedt reported, studies have found that shift work was associated with cardiovascular disease, digestive problems, lack of appetite, diabetes, and complications of pregnancy (preterm birth, spontaneous abortion, and low birth weight). It also is associated with increased accident risk; night driving greatly raises the risk of fatal collisions. At least ¾ of the shift-working population, Dr. Akerstedt found, was affected by disturbed sleep and wakefulness. He noted the problem of involuntary sleep at work: 25% of night shift process operators who can sit down while watching screens fall asleep on the job and are often unaware that they have done so. Younger workers are at highest risk of falling asleep at work on the night shift, even though laboratory studies show that daytime sleep becomes more difficult with increasing age.

Dr. Akerstedt said that the literature is inconclusive concerning whether shift work is associated with depression, cancer, or sickness absence. Night shift workers have elevated levels of cortisol and thyroid stimulating hormone. He noted that people who are worried about getting up early enough have less slow wave sleep, so that their sleep is of lower quality. These problems may affect both safety and health among truck drivers, but more research is needed in this area.

---

[*] The report on the presentation by Dr. June Fisher is out of conference sequence and grouped with social and behavioral consequences rather than with sleep disorders, because her presentation emphasized the effects of stress on transit bus drivers. It can be found in the section on "Health Consequences of Job Stress" below.

[†] His written summary of these remarks, including extensive references to the scholarly literature, is included on the accompanying CD-ROM and at *http://www.ilir.umich.edu/TIBP/truckdriverOSH*.

Dr. Ann Williamson of the University of New South Wales made a conference presentation on work/rest schedules, driving arrangements, and fatigue in long distance road transport. She presented Australian data on heavy truck crashes, showing that 49% involved fatigue, 48% involved speeding, and 14% involved illegal alcohol intoxication. The high percentage of heavy truck crashes occurring at night reflects driving patterns: truck drivers tend to drive then because the roads are less crowded at night.

Dr. Williamson reported results of Australian surveys of bus drivers and of long distance truck drivers regarding fatigue. Drivers said that fatigue affected driving in several ways:

- Slower to react (almost half agreed)
- Poorer gear changes (more than 40% agreed)
- Driving too slowly (more than one-third)
- Poorer steering (more than one-third)
- Less aware of other traffic (more than one-quarter)

Strategies used by more than 75% of drivers to deal with fatigue included listening to music or the radio, ventilation, caffeine drinks, walking or kicking their tires; but fewer than a quarter found these strategies helpful. A strategy used by only about a third of drivers, stay-awake drugs, was the only strategy rated as helpful by over half of the drivers.

More than 75% of drivers tried sleep as a remedy for fatigue, but, surprisingly, fewer than half found sleep helpful. Williamson's surveys suggested that drivers often were able to take only short naps when they needed extended sleep. Extended sleep, however, is impractical when drivers are on tight schedules that they do not control.

The survey results indicated that flexibility in scheduling rest is important in avoiding fatigue. Staged driving (called relay or "meet-and-turn" in the U.S.), in which drivers from two different locations meet halfway, exchange loads, and return to their points of origin, requires that each driver coordinate with the other. The lack of control in scheduling rest breaks when needed aggravates the problem of driver fatigue in staged driving.

In addition to surveys, Dr. Williamson conducted field studies of different driving arrangements, assessing their impact on driver fatigue and performance. One study by Williamson, Feyer, and Friswell required drivers to do three trips between Sydney and Melbourne (a distance of 900 km).[103] The first trip used staged driving; the second was driving solo and following standard driving hours regulations; and the third was driving solo but exempt from these regulations, so that drivers had flexibility to take rest breaks whenever they wished. For each trip, Dr. Williamson and her colleagues continuously measured driving performance (speed and steering variability) and heart rate. They also assessed fatigue with four off-road tests given at the beginning and end of each trip, measuring perceptual sensitivity, reaction time, vigilance, and tracking an unpredictable stimulus. Drivers experienced slightly more fatigue on staged trips. Pre-trip fatigue had an important impact on fatigue during the trip.

In another field study, Dr. Williamson and colleagues assessed driver fatigue on very long trips (4,500 km or more) in Western Australia.[104] They compared solo driving to two-up driving (called "team driving" in the U.S.), in which a pair of drivers alternate, one resting in the sleeper berth of the moving truck while the other drives. They took the same fatigue measurements as in the staged driving study. Although two-up drivers had more break time, solo drivers had longer sleep periods and were more likely to sleep at night. Two-up drivers tended to be more fatigued and to perform worse than the solo drivers. Williamson *et al.* found that a long (at least overnight) stationary rest was the best way to manage fatigue. She noted at the conference that professional drivers can manage one drive of 16 hours well; but a 16-hour drive, followed by a 6 hour rest, followed by a second 16-hour drive leads to performance problems on the second 16-hour drive.

Dr. Williamson and colleagues also conducted a laboratory study comparing the impacts of fatigue and alcohol intoxication on performance.[87,88] This study compared 20 long-haul truck drivers to 19 people not employed as drivers. Subjects were given several performance tests after 28 hours of sleep deprivation. The same subjects were given these performance tests on another occasion without sleep deprivation but with varying doses of alcohol. Williamson *et al.* compared performance deterioration associated with fatigue to the performance deterioration associated with alcohol intoxication at a BAC of 0.05%. Based on this comparison, they found that, "at the end of periods of waking of 17-19 hours, performance levels were low enough to be accepted in many countries as incompatible with safe driving."[105] Arnedt *et al.* reported similar findings.[89]

Dr. Williamson's conference presentation asserted a "need for models of fatigue management programmes that have demonstrated effectiveness in handling the compromise between operational requirements and fatigue management." She called for future investigations of the effect of breaking up long breaks into two shorter breaks and of the differences between day and night driving. Dr. Williamson said that researchers also "need validated and sensitive measures of fatigue to demonstrate effective fatigue management."[*,†]

---

[*] Ann Williamson, PowerPoint presentation at April 24-25, 2003 conference, included in on the accompanying CD-ROM and at *http://www.ilir.umich.edu/TIBP/truckdriverOSH*.

[†] A paper not addressed at the conference noted an important requirement for sensors that can measure fatigue: they must be "able to distinguish the impairment 'signal' against a highly variable 'noise'" from the driving environment." Stephen H. Fairclough and Robert Graham, "Impairment of Driving Performance Caused by Sleep Deprivation or Alcohol: A Comparative Study," *Human Factors*, Vol. 41, No. 1, March 1999, pp. 118-128 at pp. 118-119.

Dr. Drew Dawson (University of Southern Australia) gave a presentation on regulating fatigue-related risk in transport. He presented a diagram with four quadrants:

Safe and Permitted	Unsafe and Permitted
Safe and Not Permitted	Unsafe and Not Permitted

He argued that it is unwise for regulators to permit unsafe practices, and is also unwise for regulators not to permit practices that are safe. Instead, Dr. Dawson said, regulations should be tailored so that only the top left and lower right quadrants apply: all safe practices are permitted, and all unsafe practices are not permitted. He suggested a deal that would allow regulatory improvements. Regulators would ban practices that are unsafe and now permitted, and in exchange, regulators would allow practices that are safe but not now permitted.

Dr. Dawson said that regulators have tried to manage fatigue indirectly, by limiting work hours. He argued that regulators should focus on the opportunity to obtain sufficient sleep rather than limiting work hours.[*] Furthermore, Dr. Dawson argued that regulators should focus on reducing the *risks* associated with fatigue and not just reducing fatigue. Risk assessment must consider both the likelihood of a performance failure and the consequences of a performance failure. Risk may be low, he asserted, for employees doing office work or training, moderate for a team of drivers transporting innocuous cargo in a rural area, but high for a single driver transporting hazardous material through a densely populated area.

Dr. Dawson proposed offering motor carriers a choice of two regulatory systems. First, they could choose a traditional system of HOS regulation, in which driver hours would be regulated. This might be appropriate for small carriers and owner-operators. Alternatively, they could choose a fatigue risk-management system that would provide more flexibility than traditional HOS regulation but would also entail a greater information-reporting burden.

Dr. Dawson's fatigue risk-management proposal, which has been implemented in Australia for railroads and aviation, requires employers to designate an executive who can be held accountable for safety problems that are reasonably foreseeable. Carriers would educate their employees about their responsibilities to ensure that drivers have an adequate opportunity for sleep. Carriers would establish an auditable, quantitative methodology to ensure that employees have obtained sufficient sleep to operate safely.

---

[*] Validation of Dr. Dawson's strategy of focusing on the opportunity to obtain sufficient sleep, rather than limiting work hours, is a potential topic for future research. The FMCSA focused on sleep rather than on working time in its 2003 regulatory changes.

Dr. Dawson stressed the need for shared responsibility. As his presentation (included on the accompanying CD-ROM and at *http://www.ilir.umich.edu/TIBP/truckdriverOSH*) states, "The employer is responsible for providing staff with a shift system that permits sufficient opportunity to rest and recover. In determining this, the employer must acknowledge the non-work activities and responsibilities of the employee. . . . The employee is responsible for using their allocated time off to obtain sufficient sleep in order to work safely. If this has not been possible, as a condition of employment, they must notify their employer that they may have had insufficient sleep."

Dr. Dawson proposed quantitative measures to ensure that drivers have had sufficient sleep. These measures would include a "start rule" stating that a driver is unfit to work if he or she has obtained less than 5 hours of sleep in the 24 hours prior to commencing work, or less than 12 hours of sleep in the 48 hours prior to commencing work. In addition, the "finish rule" would specify, "The period from wake-up to the end of shift should not exceed the amount of sleep obtained in the 48 hours prior to commencing the shift."

Under this proposal, if either the start rule or the finish rule were violated, the driver would be required to notify the line manager. The dispatcher could then assign the driver to a late start (so that he or she could obtain additional sleep), an early finish, or an alternative job assignment where the risks from fatigue are less serious. Alternatively, the driver might take countermeasures such as napping or ingesting caffeine, or take sick leave. If, however, the employee failed to notify his line manager about the violation of the start or finish rule, then the employee would assume at least partial responsibility for any fatigue-related accidents, reducing the liability of the carrier. A similar system is used in Australian rail, where dispatchers ask employees how much sleep they have had in the past 24 and 48 hours prior to selecting their assignment.

Dr. Dawson stressed the need for the fatigue risk-management system to be simple to understand and based on readily measurable evidence (e.g., hours of sleep in the past 24 hours, rather than the driver's circadian phase) so that employees have a sense of engagement and ownership. Quantifiable measures must be available quickly enough to meet operational needs. The fatigue risk management system can employ two distinct strategies: altering shifts in a manner that reduces fatigue, and changing operations so that fatigue is less likely to result in poor job performance.

In the discussion that followed, an analogy was drawn between sleep testing and drug testing. A requirement that workers certify sleep away from work could be perceived as an invasion of privacy. Still, we already have drug and alcohol testing for transport workers; reports on fatigue are equally relevant to safety and no more intrusive. Another concern was driver cooperation in reporting sleep; a driver who repeatedly reported insufficient sleep and the need for either an alternative assignment or sick leave might be subject to disciplinary action.

Dr. Dawson was asked whether the alternative fatigue management system he proposed was feasible for a small company. He answered that a trade association for Australian aviation companies has developed a standardized fatigue management system that Australian regulators have approved and that small aviation maintenance firms just adopt this off-the-shelf plan.

One conference participant suggested a study of whether use of caffeine avoids the deterioration of driving performance associated with fatigue. Dr. van Dongen suggested a similar study involving the use of Provigil (modafinil), a wakefulness-promoting drug used to treat narcolepsy. Dr. Audrey Newell (Oakwood Healthcare System) questioned whether it was appropriate to prescribe a narcolepsy drug to patients without any medical problem; rather, work schedules should be redesigned so that healthy humans can tolerate them.

A number of participants asserted at the conference that shipper demands contribute to driver fatigue problems that affect safety. One participant suggested doing a field experiment in which shippers change their practices to ensure that drivers receive adequate rest. Such an experiment would probably have to provide financial compensation to the shippers for the extra costs that they incur.

Causes of Driver Fatigue and Sleep Deprivation

Fatigue and sleep deprivation are common problems for truck drivers. Many long-haul truck drivers face chronic partial sleep deprivation. Mitler *et al.* studied eighty long-haul drivers over a five-day period, finding that their electrophysiologically verified sleep averaged only 4.78 hours per day—and only 3.83 hours of sleep per day for those drivers on a steady night schedule.[106] While research on fatigue has emphasized long-haul drivers, fatigue also affects local/short haul truck drivers. Hanowski *et al.* monitored 42 local/short haul drivers for approximately two weeks each with video cameras and sensors, finding evidence of driver fatigue such as driving for periods with eyes 80%-100% closed.[107]

What causes driver fatigue? Brown identified three aspects of working arrangements that determine fatigue: "(a) the length of continuous work spells and daily duty periods; (b) the lengths of time away from work that are available for rest and for continuous sleep; and (c) the arrangement of duty, rest, and sleep periods within the 24-h cycle of daylight and darkness, which normally entrains individuals' circadian rhythms."[108] Truck drivers, Brown asserted, are particularly at risk for fatigue because their work schedules are irregular and beyond the driver's personal control; their sleep breaks often occur during the day when conditions are not favorable for sleep; they experience stresses in the truck cab such as heat, noise, and vibration; and they continue working even when fatigued in order to reach their destination. The latter is consistent with Beilock's finding that schedule pressures often induced truck drivers in Florida to violate HOS regulations or speed limits, particularly for drivers who were driving solo, had long trip distances, or refrigerated loads,[109] as well as with Williamson's findings discussed above. Arnold et al. surveyed truck drivers and motor carrier managers in Australia regarding causes of driver fatigue, finding that the cause most often cited by both groups was driving long hours.[70]

Akerstedt and colleagues have authored several studies on work schedules, sleep, and fatigue. An EEG study by Torsvall and Akerstedt showed that ships' engineers who worried about being awakened while they were on call did not sleep well.[110] Truck drivers are commonly in this situation, having sleep breaks while awaiting telephone notification at an unpredictable time informing them of the availability of their next load (a situation explicitly permitted by the current HOS regulations). Similarly, a survey by Akerstedt et al. of over 5,000 employed adults in Sweden found that problems unwinding after work often disturbed sleep.[111] At the conference, Dr. Akerstedt said that the strongest correlates of fatigue were night work, a long time awake, and backwards rotation of shifts; day sleep and early shift starting times had weaker correlations with fatigue. It was unclear, Dr. Akerstedt said, whether on-demand scheduling led to fatigue.

A study by Cziesler *et al.* noted the importance of circadian principles in designing work schedules.[112] They studied 85 industrial shift workers who initially had a weekly backward shift rotation (an 8-hour phase advance, in which workers' start times rotate from day to night to afternoon). Workers were happier, and their productivity was higher, when their shift rotation was changed to a forward rotation (an 8-hour phase delay, in which workers' start times rotate from day to afternoon to night) once every 21 days.

Monk identified a problem with slow rotation (such as every 21 days) or steady night shifts: workers revert from a nocturnal schedule to a diurnal schedule during the weekend in order to spend time with family and friends, undermining the adaptation to night work.[113] But Knauth[114] concurred with Cziesler et al. that forward rotation was better for workers than backward rotation.

Truck driver work schedules, however, often are not in accord with the Czeisler et al. recommendation. As Saltzman and Belzer noted, a 1962 amendment to the federal HOS regulations for interstate truck drivers permitted a driver in a hurry—or a carrier dispatching a driver for maximum productivity—to adopt an 18-hour cycle: 10 hours of driving, followed by eight hours off-duty, followed by another driving cycle.[37] This is a substantial backward rotation of up to 6 hours per day compared with 8 hours per week reported in the Czeisler et al. study. As noted above, even the new HOS regulations have within them the potential for substantial backward or forward schedule rotation, if the driver does not work or drive the maximum possible hours. In the example in Tables 4 and 5, assuming the driver drives all available hours and does not use any time for non-driving labor, the driver experiences a three-hour backward rotation with the new regulations. If the work pattern requires even shorter days, the backward rotation can become even more pronounced. All of the predictable options made for the HOS regulations, therefore, go "out the window" as drivers' true practice reflects the often quite irregular demand for trucking services.

This irregular nature of driver work/rest cycles presents an additional systemic problem. Truck drivers often must work or attempt to sleep at an inappropriate circadian phase. If a person's work/rest cycle is synchronized with the solar day, the endogenous circadian signal helps the individual maintain alertness for a full 16 hours. In the early evening homeostatic sleep pressure – the need for restorative sleep after a long time awake – begins to build, and eventually helps the individual fall asleep. At 3 AM, after the first few hours of sleep have relieved the homeostatic pressure, the circadian signal helps the individual stay asleep. Similarly, if a person whose circadian clock has been entrained to the solar day tries to work all night, homeostatic pressure and the circadian signal will combine to make him or her sleepy.[115] Furthermore, sleeping during the day may be difficult, even if he or she has been awake for a long time.

One conference participant said that regional freight generally is loaded early evening, and the driver is required to deliver it by the following morning. In these circumstances, when the carrier's operations have a daily pattern of local freight pickup and overnight delivery (as in the regional LTL market), city drivers will work during the day while road drivers will run through the night. These drivers' problems arise paradoxically on and following the weekend, when they shift to their families' normal day shift and back again to the night shift the next week.

Several conference participants cited the need for further study regarding how much sleep truck drivers actually get. It is well known, however, that drivers often do not accurately record sleep in diaries, so it might be better to use an actigraph to register hand movements indicating that the driver is awake. Drivers, however, can manipulate actigraph measurements; even EEG measurements can be manipulated by putting the electrodes on a person other than the one the researchers intended to study.

Health Consequences of Driver Work Schedules

In 1908, a brief by Louis Brandeis persuaded a property-rights-oriented U.S. Supreme Court to uphold the constitutionality of an Oregon statute limiting women's work to ten hours per day, on the grounds that longer work hours could injure the health of women workers.[116] While at the time it was controversial even to limit women's hours, proponents could justify it based on protective concerns. We now know that long and irregular work hours have an adverse effect on the health of workers in some occupations, notably including truck driving, regardless of gender. Saltzman and Belzer argued that improved driver health was potentially a large benefit of stricter federal regulation of driver work hours.[37] The report by Hamelin[55] (included on the accompanying CD-ROM and at *http://www.ilir.umich.edu/TIBP/truckdriverOSH*) notes that high driver turnover makes a portion of the costs of poor driver health an external cost from the perspective of motor carriers, which implies that free markets provide less than the efficient level of driver health.

Truck drivers may not fully recognize the harmful effects of their work schedules on their lives until the long-term effects have become evident. Spelten, Barton, and Folkard examined the extent to which shift workers underestimate the harmful impact of their jobs on their health and well-being.[117] They surveyed retired British police officers who had experienced shift work prior to retirement (typically at age 50) and then taken on new jobs, generally on the day shift. They also surveyed current British police officers. The retirees gave two measures of their health, fatigue, and anxiety levels while they had been police officers: their recollection of how they would have scored them at the time, and the score that they now thought was accurate. Their "at the time" scores reported fewer problems than did their post-retirement assessments. Nevertheless, their "at the time" scores were similar to the scores reported by currently employed police officers. Spelten *et al.* concluded that shift workers lose sight "of what normal (i.e., day-work) life is like. That is, they may habituate to a gradual lowering of their well being, subjective health, etc., and hence consider (and thus rate) themselves as relatively 'normal.'"[118]

As Saltzman and Belzer reported,[37] chronic partial sleep deprivation is a major cause of driver health problems. Spiegel *et al.* found that restricting sleep in healthy young men to four hours per night for a mere six nights "is associated with striking alterations in metabolic and endocrine function." Specifically, sleep debt reduced glucose tolerance and thyrotropin concentrations, and it increased evening cortisol concentrations and the activity of the sympathetic nervous system. "The effects are similar to those seen in normal ageing and, therefore, sleep debt may increase the severity of age-related chronic disorders" such as diabetes and hypertension;[96] see also Leproult *et al.*[119] This hypertension may account for the high risk of stroke found among truck drivers in Denmark.[120]

Other researchers have confirmed the harmful effects of partial sleep deprivation on healthy, working-age men. One study assessed the impact on blood pressure of overtime work that limited sleep. Blood pressure was significantly higher following a day with overtime work and only three to four hours of sleep than it was following an eight-hour workday and approximately eight hours of sleep.[121] Another study compared immune function after a normal night of sleep to that after a night when subjects were not allowed to sleep between 10 p.m. and 3 a.m. The researchers found that "even a modest disturbance of sleep produces a reduction in natural immune responses," resulting in increased vulnerability to infection.[122] A third study measured sympathetic nervous system activity, both on nights when subjects were allowed to sleep and on nights when subjects were awakened at 3 a.m. and kept awake until 6 a.m. They found that partial sleep deprivation raised nocturnal catecholamine levels, which can contribute to cardiovascular disease.[123] This laboratory finding was supported by epidemiological evidence: middle-aged men who suffered sleep loss as a result of rotating shifts had higher risks of coronary heart disease than men working only during the day.[124]

Working long or irregular hours may have other harmful effects on health in addition to those related to partial sleep deprivation. For reviews of the literature, see Michie and Cockcroft,[125] Spurgeon *et al.*,[126] Harrington,[127] and Sparks *et al.*[128] An epidemiological study in Sweden examined the impact of long work hours on mortality between 1973 and 1996 among approximately 11,000 men and 9,500 women born between 1926 and 1958. Even controlling for age and for behavioral factors such as smoking, drinking, and use of tranquilizers, regular overtime of more than five hours a week was associated with higher mortality rates for five years following the overtime.[129] Knutsson noted the lack of conclusive evidence that shift work shortens life spans or increases the risk of cancer; but his literature review found some evidence that shift work increased the risk of diabetes and metabolic disturbances and strong evidence that shift work increased the risk of coronary heart disease, complications of pregnancy, and peptic ulcers.[130]

Other studies have examined the impact of long or irregular work hours on specific health problems:

- Extended working periods desynchronize the internal circadian rhythms of long-haul drivers who work many hours per day and have work/rest cycles less than twenty-four hours.[131]
- Irregular hours and night work raise the risk of being hospitalized for ischemic heart disease (IHD).[132] Professional drivers are at greater risk of IHD if they work long hours.[133] This cardiac risk may increase partly because professional drivers who spend long hours behind the wheel tend to have a higher body mass index.[134]
- Working over 40 hours per week doubled the risk of acute Helicobacter pylori infection (associated with peptic ulcers), even controlling for age, sex, and marital status.[135]
- A group of Dutch truck drivers working an average of 11.4 hours per day had insufficient recovery after work from sympathoadrenal activation. Their elevated catecholamine levels were associated with increased psychosomatic health complaints.[136, 137]

Apropos of this research, Dr. June Fisher commented that there is an extraordinarily high level of cardiovascular disease among bus drivers in Scandinavia and Germany. She suggested that the reason might be that weekly shift rotations are common practice there (perhaps rooted in an egalitarian desire to share the burden of working night shifts).

Health Consequences of Job Stress

Truck drivers experience stress for several reasons. Long hours and extended periods away from home cut them off from friends and family; and even on days home from work, they may be too tired to nourish their marriages or their relationships with their children. Unless their home lives are exceptionally stressful, they likely will enjoy their off-duty hours less if they are spent in the sleeper berth of a truck rather than at home. In fact, regularly being "stuck" on the road in a strange city puts intolerable stresses on many—perhaps most—average people, and truck drivers may be stranded in undesirable locations as well. Owner-operators often are under intense financial pressure, finding it difficult to make the required loan or lease payments on their truck. Package drivers may experience stress because of intense work—a combination of many stops per day and pressure to pick up and deliver on time. A study of package drivers in the U.S., for example, found that their mean score on a standard scale of psychological stress was at the 91st percentile for the general adult population.[138] Truck drivers also may experience stress from dealing with customers to whom they deliver loads: they often are required to wait in their trucks for long and unpredictable periods of time before being brought in to the dock to load or unload; are denied opportunities for food, water, and restroom facilities; and are treated disrespectfully by shipping and receiving personnel.

Some studies of job stress among drivers have focused on bus rather than truck drivers. For example, Raggatt and Morrissey found that physiological measures of heart rate, blood pressure, catecholamines, and cortisol confirmed that long-distance bus drivers experience stress during long work shifts.[139] Sluiter *et al.* found that this stress response may be aggravated if long-distance bus drivers do not have adequate resting times during trips and a duty-free recovery period between trips.[140]

Bus and truck drivers have somewhat different job duties; for example, dealing with passengers is an important job duty for bus drivers, while loading and unloading cargo is an important job duty for many truck drivers. Employment conditions in the urban transit segment of bus transportation may be affected by stark economic differences with trucking: urban transit drivers work for subsidized, government-owned monopolies and are often protected by union contracts, whereas truck drivers work for private companies facing intense market competition and, in the TL sector, are rarely unionized. Urban transit drivers also are not subject to the same federal HOS regulations as are interstate truck or bus drivers. Still, both urban transit and long-distance bus drivers encounter some of the same problems as truck drivers do, such as whole body vibration and difficulty unwinding after work. Long-distance bus drivers, like regional and long-haul truck drivers, are often away from home and work at times when their circadian clock is set for sleep, and they often have irregular work schedules. While the differences between bus driving and truck driving are important, research about stress among bus drivers may nevertheless have some relevance for truck drivers.

At the conference, Dr. June Fisher (University of California-San Francisco) presented research she conducted with Dr. David Ragland of the University of California-Berkeley on how job stress affects the health of urban transit operators (including not only bus drivers, but also trolley and cable car drivers and light rail workers). Dr. Fisher directs the MUNI Health and Safety Study, a large study of workers in San Francisco's MUNI urban transit system. This study began in 1978 as a result of concern about occupational health by Transport Workers Union Local 250A, which represents MUNI employees. Researchers conduct biennial medical exams mandated for urban transit drivers by the U.S. Department of Transportation, and they use the exam results to analyze health trends. Two important features of the study are that, first, providers conducting the medical exams have knowledge about the work environment and, second, researchers have ongoing interactions with both MUNI management and the union.

Dr. Fisher reported that the MUNI study found a high rate of hypertension among urban transit workers, which she attributed to stress. The separation rate was higher for hypertensive MUNI workers than for non-hypertensive workers, even controlling for age. Dr. Fisher reviewed studies conducted by others of urban transit operators' health and safety, noting that transit operators have higher than average levels of cardiovascular and musculoskeletal risk.

Dr. Fisher said that some health behaviors of MUNI transit workers contributed to health risks. Body mass index was higher, on average, for employees with over 10 years of job tenure than for newly hired employees. Employees with five or more years of job tenure with MUNI smoked more cigarettes per day and consumed many more alcoholic drinks per week than did newly hired employees. Dr. Fisher raised the question of whether these behaviors were maladaptive coping mechanisms for dealing with job stress.

Dr. Fisher showed a diagram about occupational stress, first presented to MUNI transit operators in 1986 by the late Professor Bertil Gardell. Stressors include both overload (e.g., time pressure or more task complexity than the operator is prepared to handle) and "underload" (e.g., too simple a job role relative to the operator's capabilities). Other stressors include ambiguous job instructions and awkward work postures. Stress can be affected, Gardell's diagram indicated, by decision latitude regarding one's own work (e.g., control over job content, work methods, and pace, and physical freedom); social support from supervisors, family, and friends; and control over future prospects (e.g., information about major changes at work and access to continuing education).

Dr. Fisher reported that drivers who face substantial time pressure (because they get behind the timetable and must strain to get back on schedule) exhibit physiological stress, as indicated by their cortisol and adrenaline levels. Such drivers take more frequent sick leaves and more often report stomach pains. Other occupational stressors in urban transit besides time pressure include violence, troublesome passengers, traffic, lack of supervisory support, shift work, and long work hours. The consequences of transit operator health problems for the MUNI urban transit system, Dr. Fisher continued, include absenteeism, increased workers' compensation costs, increased turnover, and early disability retirement.

Dr. Fisher recommended that future research include work-based fatigue studies, physiological and biochemical monitoring studies at work and rest, assessment of the impact of rest periods on vigilance and fatigue, studies of unwinding, development of methods to reduce maladaptive coping strategies, and studies of how changes in the work environment affect hypertension.

A potential topic for future research is to determine the extent to which research findings concerning job stress among long-distance bus drivers and urban transit workers are applicable to truck drivers.

Social and Behavioral Consequences of Employment as a Truck Driver

A review by Costa noted the impact of nonstandard work schedules on social relationships:[141]

> People engaged in shift- and night work are frequently out of phase with society, and can face greater difficulties in their social lives because most family and social activities are arranged according to the day-oriented rhythms of the general population. Consequently, shift work can lead to social marginalization due to the mismatch between the shift workers' time budgets (working hours, commuting, and leisure times) and the complex organization of social activities.

Truck driving jobs, especially in long-distance trucking, entail not only long and irregular work hours, often at night, but also frequent travel away from home and pressure to meet tight deadlines. Mr. Ron Jager, an owner-operator, noted at the conference that living in your truck is like living in an RV—only you don't get to choose where you go. These difficult job characteristics may have adverse social, behavioral, and other consequences for truck drivers and their families.

Dr. Lawrence Root (University of Michigan School of Social Work and Institute of Labor and Industrial Relations) presented qualitative data on the effects of long work hours on work and family issues. In the U.S. labor force as a whole, most of the employees taking family leave or requesting flexible hours are women. He noted, however, that long work hours for men also had a substantial impact on the family. He presented statements about work hours and family from three male blue-collar workers, interviewed by Dr. Root's University of Michigan colleagues:

> . . . the thing that I remember the most was that my father used to work seven days a week. Even though we had all the material things that we could ever want, he was hardly ever there. . . He was always at work. That's what I remember the most about my father growing up. . . . I don't want to have the same relationship with my son—I don't want my son to remember me as someone who worked all the time.
>
> —Harold, a new employee at "Sylvania"

> I would work every overtime there was to cover for the fact that my wife didn't work outside the house. But like I said she was there to raise the kids. I know whenever they went to school they had some food in they stomachs and that they had the right clothes on. And I know when my girls went to school they wasn't in elementary with a face full of make-up on. Because she was there to see them walk out that door and to see them when they came back in that door. But a lot of people wasn't that fortunate. Kids leave out and they have maybe another thing of clothes in

their backpack. But my wife would make sure that when you leave outta there, "Let me see what you got in that bag." You know what I'm saying. (laughs) So that all paid off.

—Arthur, 30-year worker at "Sylvania"

[Overtime and shift-work] got me away from them more than I should have 'cause I, I worked long hours. When I worked midnights, one of the reasons I worked midnights is made more money. I did it for 3½ years. I could go in like when they were goin' to bed. And I you know be home and asleep when they got home from school I was there. And I liked that. My wife hated it and I didn't realize how much she hated it. That almost caused us to be divorced. . . . I thought, "Yeah you're complainin' ya got kids, ya got 2 dogs, ya got a nice house. Why don't ya—." Not being very sympathetic. That's why I say she put up with more than I realized. . . . She said, "You don't get off midnights, you're not gonna have anyone ta come home to." And that's when I knew she was serious. (laughs) . . . I tried not to let work interfere with my kids. And I think sometimes I let it interfere with my wife. 'Cause the kids were more of the priority and I just always thought the wife would understand. And I learned like I saw about that midnights and that she didn't understand. She didn't understand why she wasn't being more a priority. (pause) Work came first on Saturday.

—Don, approaching retirement at "Sylvania"

Dr. Root also presented data from the University of Michigan Trucking Industry Program (UMTIP) driver survey in 1997 and 1998, which included over-the-road truck drivers who were employees as well as owner-operators. Ninety-seven percent of the respondents were male, 61% were married, and 57% had children. Between the last trip and the trip during which they were interviewed, 52% had slept at home, while 42% had slept in their truck. Twenty-six percent said they were "not at all satisfied" with the amount of time they have at home, while 21% said that their company was "not at all concerned" with the need for time at home and the well being of the families of truck drivers in their employ.

Dr. Root noted several challenges for work-family balance with truck drivers. First, the stress of long work hours for truck drivers is compounded by often being far from home. Second, work-family discussions often have provided little attention to truck drivers and other predominantly male occupations because of the traditional view that work-family balance is mainly of concern to women. Third, programs such as family leave and flexible scheduling may be more difficult to implement in the context of trucking. Finally, basic improvements to make trucking more "family friendly" require looking more fundamentally at the structure of the work relationship.

In the discussion at the conference, several people addressed social relationships of truck drivers. Dr. Dawson reported on a study he did of Australian rail workers, who seemed to over-report time that they spent with their children. He said that a successful shift worker in Australian railroads often had traditional marriage with a stay-at-home wife who did everything for him at home. Dr. Quinlan said that truck drivers have a terrible work/family balance, but there is relatively little research on this. Dr. Newell noted that the wife of a trucker is like a single mom, assuming almost all of the responsibility for taking care of the children and the home. The U.S. military provides support for the spouses of soldiers who recently went to Iraq. Could something similar, Dr. Newell asked, be provided for spouses of truck drivers? Dr. Dawson replied that Australian rail workers have support groups for spouses and a "lend a husband" program for home repairs.

Dr. Newell suggested a study of the impact of trucking employment on relationships with spouses and children. Do the children of long-haul truck drivers lack parental help with homework and have difficulty in school? Does employment as a long-haul truck driver increase the likelihood of divorce? Alternatively, do frequent absences from the home provide a respite from family conflict and reduce the likelihood of divorce? The latter could be the case if long-haul truck drivers are disproportionately likely to have poor interpersonal skills.

Dr. Fisher noted the harmful behavioral consequences of employment as an urban bus driver. Such drivers face conflicting role demands. They need to maintain constant vigilance to maintain safety. They face pressure to stay on schedule. They are required to be nice to passengers, even troublesome passengers. Bus drivers deal with their tensions by smoking and drinking. They do not drink on the job, but they drink more after work once they start working as drivers. Dr. Fisher saw a need to study methods to reduce maladaptive coping strategies among bus drivers.

Truck drivers, too, may employ maladaptive coping strategies and behaviors related to a life on the road. Further research could determine if employment as a truck driver leads to increased smoking, poor diet, lack of exercise, and an increased incidence of sexual relations with prostitutes (a practice that contributed to the spread of AIDS in Africa). More importantly, future research could evaluate potential countermeasures, such as the wellness-counseling program for professional drivers evaluated by Hedberg *et al*.[142]

When John Siebert said that OOIDA members did not have unusually high health insurance costs, Dr. Newell was puzzled. Since truck drivers have higher rates of diabetes, hypertension, and other chronic diseases, their use of medical services should be higher as well; and these conditions are expensive to treat. She suggested that truck drivers may not be getting needed medical care because of scheduling problems and being out of town.

Dr. Newell's suggestion at the conference was confirmed by a subsequently published study by Solomon *et al*.[143] This study found that 47% of long-distance truck drivers surveyed in 2002 lacked a regular health care provider. Fifty-six percent of drivers "reported difficulty making a healthcare appointment while at home . . . due to their work schedule. . . .[and 62%] needed healthcare. . . but did not seek it because they were on the road working."[144] Future research could explore ways of organizing the provision of medical care so that truck drivers have access to health services, including continuity of care needed for effective diagnosis and treatment.

Developing a Research Agenda on Truck Driver Occupational Safety and Health

Near the end of the conference, participants broke up into discussion groups organized by area of interest. The following section includes a literature review on issues raised at the conference and issues that have arisen in discussions with interested parties since the conference. It also includes the reflections of the conference participants regarding what they had learned and where research should be directed.

⇒ What countermeasures might be taken to address driver fatigue?

A review article by Rosa et al. listed several possible interventions to promote adjustment to night work and shift work:[145]

- Redesigning work schedules (e.g., forward rotation of shifts rather than backward rotation)
- Rest periods or scheduled naps within shifts to relieve fatigue
- Exposure to bright light to reset circadian clocks
- Physical activity and conditioning
- Pharmacological aids
- Eating habits conducive to sound sleep hygiene
- Individual behavioral techniques (e.g., relaxation training to reduce stress)

A recent review by Knauth and Hornberger asserted that extending work shifts beyond 8 hours leads to fatigue unless there are sufficient breaks during the shift and time after the shift for complete recovery. Employees, they continued, need at least 11 hours off between two work shifts, as rest periods of 8 to 10 hours may allow only 3 to 5 hours of sleep.[146] At the conference, Dr. Akerstedt said that shift workers were better off if they worked a few long shifts *provided* that they had frequent days off (e.g., working 12 hours two days in a row, followed by two full days off work).

⇒ How serious is the effect of irregular and unpredictable scheduling? How can policy-makers reduce the irregularity and unpredictability of scheduling without introducing potentially devastating costs into the logistics chain?

Truck drivers often face a problem that most factory shift workers do not: highly irregular and unpredictable schedules, as well as work that takes them away from home as a normal process, so whether drivers could productively take the kind of time off suggested by Akerstedt depends greatly on the nature of the work process[*]. Horowitz *et al.* found

---

[*] At the time that J.B. Hunt implemented the famous increase in compensation that was the subject of the study by Belzer, Rodriguez, and Sedo, they also offered a "get-home" policy that promised the drivers that two weeks after they asked to be home the company would get them home regardless of cost. See Belzer, Michael H., Rodriguez, Daniel A., and Sedo, Stanley A. (2002), "Paying for Safety: An Economic Analysis (Continued on next page)

that maintaining a consistent sleep schedule facilitates adaptation to night work.[147] Conference participants noted, however, that shipper cooperation is necessary to provide predictable work schedules that allow drivers to sleep well. Trucking is very demand-driven, reflecting both the power of shippers and industrial practices such as just-in-time inventory systems. Drivers often do not know when to be ready for work. A dispatcher or broker may string a driver along regarding when a load will be ready, and the driver stays awake expecting a job, but then the job may not start until 12 hours later. Dispatchers can telephone the driver during a rest break to notify him or her when to report back to work. The sleep disruption may substantially affect the driver's recovery from fatigue and eventually damage his or her health. An operational change that would reduce driver fatigue would be to notify drivers regarding the timing of their next dispatch at the end of their tour of duty. Effecting this change would require a major cultural change in the trucking and shipping community, which is accustomed to demanding service whenever the freight is ready to move, and on very short notice.

> ⇒ Will some people naturally tolerate the scheduling problems that are traditional in trucking and transportation? What personal characteristics explain differences in drivers' ability to deal with night work or irregular shifts?

At the conference, Dr. Rogers said that some people are better able than others to tolerate night work or irregular shifts. A recent review by Costa reported, "it is not yet completely clear to what extent personal characteristics influence long-term tolerance to night work and hence whether they can be used as possible predictors for such tolerance."[148] Further research to predict which individuals best tolerate driver work schedules would be useful for personnel selection.

> ⇒ What time or other intervention might help truck drivers better recover from sleep loss without harming their health?

Dr. Reid said there was a need to know more about recovery from sleep loss. Dr. Akerstedt suggested conducting field experiments on fatigue with companies using different work schedules, compensating companies for incurring higher costs associated with the study. Dr. Drake suggested an experiment in which truck drivers would be paid to sleep, assessing whether this offsets the financial incentive of mileage-based pay to get less than an adequate amount of sleep. Dr. Fisher suggested a field study of the impact of rest periods on fatigue.

> ⇒ How can drivers and other workers similarly situated better unwind after work? How can unwinding be accomplished without the use of self-medication? How can drivers be socially reintegrated after a stressful day of driving?

(Continued from previous page)

of the Effect of Compensation on Truck Driver Safety," (Washington, DC: United States Department of Transportation, Federal Motor Carrier Safety Administration), pp. 8, 75.

Dr. Fisher also saw a need for further studies of unwinding after work. Dr. Newell noted that medical interns and residents are very tired after a 36-hour work shift but still cannot get to sleep immediately because they are too wound up; drivers may face similar problems, particularly if they have been driving under severe time pressure or during adverse weather conditions. Some of the interventions mentioned by Rosa *et al.*[145] might help with unwinding. One topic for future research is the impact of biofeedback-assisted relaxation training for drivers on their ability to get to sleep promptly after completing a work shift. It would also be useful to assess the impact of exercise programs on the quantity and quality of driver sleep.

> ⇒ How might the interests of carriers and shippers be structured to align with the health interests of drivers?

A conference participant said that trucking firms with poor management practices (e.g., those that accept jobs below cost) also tend to have fatigue problems because they do not refuse loads that impose extremely demanding work schedules. Another participant suggested conducting a field experiment that tries to get shippers to change their practices so that drivers can get more rest and avoid fatigue.

> ⇒ What are the long-term health implications of addressing drivers' inability to adjust to irregular shift schedules by using pharmacologic or other approaches that force or trick the body to conform to commercial scheduling demands?

Some studies have examined the use of benzodiazepines[149] and melatonin[150] to address sleep problems associated with the circadian clock. Van Cauter and Turek, however, argued that "pharmacologic approaches may be acceptable for the short-term treatment of jet lag but not for the long-term management of shift work."[151] Although these drugs may achieve a safety goal by tricking the body in the short term, regular use to overcome fatigue or sleepiness may have the long run effect of damaging the driver's health.

Another approach for the chronic sleep problems of truck drivers with irregular work schedules might be that suggested by Czeisler et *al.*: using exposure to bright light to increase alertness during night work shifts, and nearly complete darkness during daytime rest periods to facilitate sleep.[152] In ten two-week studies of eight men, they found that the light/darkness treatment increased daytime sleep in night shift workers by an average of two hours per day. The Czeisler *et al.* experiment entailed exposure to bright light throughout the night shift—something that is not feasible for truck drivers working at night.

A potential future research project might assess the impact on fatigue of equipping sleeper berths of trucks with bright lights (~10,000 lux) and having drivers use eyeshades if they sleep during the day in motels with insufficient light-blocking drapes. Drivers who needed to drive during the night and sleep during the day could reset their circadian clocks by brief exposure to bright light at night and complete darkness during their rest periods. Other aspects of sleeper berth design, such as more extensive soundproofing, also may merit study.

⇒ Should additional public or private highway rest areas be created to provide drivers with a safe place to stop and sleep?

A potential topic for future research not addressed at the conference is the extent to which construction of additional highway rest areas or expansion of parking facilities at existing rest areas would improve highway safety. This has been the subject of numerous studies by transportation researchers, whose policy recommendations often have been blunted by truck stop operators opposed to public solutions to what they see is a private problem. For an example of such research, see FHWA (2002)[153]

⇒ How do we design and implement successful remediation programs in support of healthier, safer truck drivers?

Success in improving truck driver occupational safety and health depends not only on scientific understanding of linkages between employment and health, but also on effective implementation of remedial programs. It is therefore important to conduct research on how to disseminate knowledge of best practices and how to secure the cooperation of the relevant parties—employee drivers, owner-operators, for-hire carriers, private carriers, unions, shippers, consignees, and others—in adopting these practices.

⇒ Field studies appear to be a necessary component of a successful research agenda. Can researchers gain access to the affected private-sector firms?

Field studies require the involvement of affected parties. At the conference, Mr. Madar of the Teamsters said that his union was willing to participate in field studies related to driver occupational safety and health, but that they would want to be consulted at the outset—not presented with a *fait accompli* negotiated between researchers and trucking companies. Fortunately, this conference included representatives of a variety of industry groups, including the Teamsters, OOIDA, the American Trucking Associations, the Motor Freight Carriers Association, the American Moving and Storage Association, and the American Insurance Association. Consultations with these groups and others in the industry are essential for good research on truck driver occupational safety and health.

⇒ How can the research community get the attention of drivers themselves? How can drivers better communicate their needs to researchers? Can collaborative research be designed that engages the trucking community so that they believe their full range of interests have been considered?

Engagement is necessary to ensure that best practices are understood and adopted. A study in the 1970s by Kochan, Dyer, and Lipsky noted the benefits of involving labor unions in efforts to improve occupational safety and health.[154] Research on truck driver occupational safety and health could similarly assess the effectiveness of cooperative labor-management programs, either in the union sector (involving parcel, LTL, or other carriers and the Teamsters) or in the nonunion sector (involving TL carriers and OOIDA, for example).

⇒ Can research validate the proposed regulatory emphasis on restorative sleep rather than working time?

This was the approach recommended by Dr. Drew Dawson, and the approach taken by the Federal Motor Carrier Safety Administration in its 2003 and 2005 regulations. Does it work?

⇒ Do safety and health pay? What would it take to reconfigure incentives in the trucking industry to align industry practices with public health goals for truck drivers?

The cooperation of individual drivers, too, is crucial to efforts to improve occupational safety and health. Drivers value their health. They also, however, value their earnings and the chance to spend non-working time at home. If wage rates are constant, shorter work hours imply lower earnings. Furthermore, strict limits on work hours sometimes require drivers to stop and take their required break in the sleeper berth of their truck, even if they are not tired and though they would prefer to continue driving until they reach home. Belzer, Rodriguez, and Sedo showed how driver compensation predicts safety both for the drivers and for the carriers for whom they work. However, they have not been able to demonstrate the extent to which the cost of this higher compensation offsets the safety and health benefits. Since a market becomes more efficient if it can internalize externalities, such as safety and health, it might make sense to consciously structure trucking regulations that line up individual and firm incentives with social objectives such as highway safety. Researchers should study the extent to which driver preferences change when they are provided with more complete information about the health and safety consequences of current driving practices.

- Would effective communication of occupational safety and health information persuade drivers to accept some sacrifice in earnings?
- Would it persuade them to accept the need to stop for a rest break if they cannot reach home before they reach limits on work hours?
- Would it persuade drivers (as well as trucking companies, shippers, and policy makers) that they deserve compensation for the health and mortality risks they take?
- Would full information on the health and safety consequences of current labor market practices persuade motor carriers, shippers, and consignees that better compensation for drivers is in their business interest?
- Would such information persuade the public that their greater safety, as well as the cost of driver health care and the cost of shorter work lives, is worth the cost of paying drivers more money and reducing their work stress?
- Would policy makers accept the notion that drivers deserve greater bargaining power to improve their earnings?
- Could dynamic programs such as safety benchmarking be used to identify characteristics of safe firms and foster the development of continuous improvement among motor carriers?

We have consolidated the full set of recommendations in the following tables. The wide range of interests, fields of study, and disciplines represented at the conference speaks to the breadth of the problem faced by these workers. Data needs were a high priority for many of the scholars who participated, because without data, research cannot begin, and without data, research has little impact.

Table 10. Research Needs: Data

⇒ The development of more systematic data on morbidity on truck driver occupational safety and health is critical for any research agenda on this population. Health and safety as well as working conditions are difficult to measure for this mobile and hard-working population.
⇒ Data must be collected on life expectancy and disability specific to truck drivers without having to rely on privately funded and collected data that may not represent population outcomes well.
⇒ Another important area for future research is to replicate with American data the foreign epidemiological studies concerning the incidence of disease among truck drivers.
⇒ Are workers' compensation statistics on occupational injuries and illnesses consistent with other data sources, such as Annual Survey of Occupational Injuries and Illnesses conducted by the Bureau of Labor Statistics?
⇒ Quantify the risk drivers face due to extended exposure to highway, wind, engine, and mechanical noise associated with trucking operations.
⇒ Field studies appear to be a necessary component of a successful research agenda. Can researchers gain access to the affected private-sector firms?
⇒ Special studies should be undertaken on the effects of team operations on truck driver occupational safety and health. Data should be collected on the nature of operations as well as the effects of real-world operations on drivers and their families.

Occupational safety and health does not exist in a vacuum. Occupations are situated within industries and they are set within a marketplace. People do not perform work in the abstract or as a hobby; they do it in trade for the money they use to pay their cost of living and to support their families. As part of the economic process, work is subject to pressures that may induce or force people to make a bigger trade than they thought they were making. For truck drivers, their work may be in trade for family life and it may be in trade for their health and longevity. It is essential, therefore, to undertake the research required to determine the extent of that tradeoff. While it is unlikely that the character of the trucking job will change fundamentally in the foreseeable future, anything scientists and policy makers can do to mitigate the damage and provide full information to everyone participating in the economic process, the more likely the outcome will be fair.

Table 11. Research Needs: Economics and Industrial Organization

⇒ How might the interests of carriers and shippers be structured to align with the health interests of drivers?
⇒ Do safety and health pay? What would it take to reconfigure incentives in the trucking industry to align industry practices with public health goals for truck drivers?

⇒ Motor carriers justify team operations because they believe they provide a cost-effective way of improving service (thus improving the competitiveness of trucking in comparison with other transportation modes).
    o How safe are team operations compared with solo operations on similar roadways?
    o Building on FMCSA's "Impact of Sleeper Berth Usage on Driver Fatigue,"[155] how frequently are the conditions recommended on the basis of focus groups and intensive study of team drivers maintained in actual operations?
    o What are the long-term health and safety effects of team driving?
    o Do the economic benefits of team operations outweigh the social, health, and safety costs?

⇒ To what extent does the higher claims cost for workers' compensation cases with attorney representation stem from attorneys selectively agreeing to represent clients who have a strong case and large potential claims?
    o To what extent does the mere involvement of an attorney affect the outcome?
    o Do attorneys win larger settlements than the claimants would have received without attorney representation, or are attorneys involved in cases more likely to have larger settlements?
    o How can we weigh the costs of allowing workers' compensation claimants to shop for doctors against the individual's right to choose his or her own medical care?

Before beginning the research agenda, it is important to assess the nature of the health and safety problem. The situation of commercial motor vehicle drivers is particularly difficult because their health and safety risks are caught up in a confounding context – one that may mask the true sources of the problem. The focus on individual and particular safety and health issues often may miss the mediating factors that are most important. To build on the example cited in the conference, if economic forces compel drivers to work such long and irregular hours that it damages their health, it will do no good to focus just on those hours of work; the economic forces that compel people to work that hard must also be considered.

Table 12. Research Needs: Assessment

⇒ Research must be done to assess the subtle linkages between employment conditions for drivers and health problems such as diabetes and cardiovascular disease because of their heavy toll in morbidity and mortality.
⇒ Epidemiological and economic analysis must be combined, helping researchers estimate of the dollar cost of morbidity and premature mortality associated with employment conditions for truck driver
⇒ Can routine truck inspections readily detect diesel emission problems that can be fixed inexpensively by such measures as cleaning or replacing the injectors and sealing off leaky gaskets that leak exhaust into the truck cab?
⇒ Is it cost effective to reduce diesel exhaust exposure by installing filters in the heater and air conditioner air intake systems of vehicles such as pickup and delivery trucks, urban buses, and taxis that are driven for extended periods in congested urban areas?
⇒ Research should be undertaken to replicate the McGlothlin study of beverage delivery drivers with other groups of truck drivers, particularly in LTL and package delivery.
⇒ Research is needed to determine the extent of occupational violence in trucking. Researchers should also measure the effects of job stress, including the stress of repeated unpredictable dispatch, conflict with dispatchers and supervisors, and the stress of isolation from family and social supports.
⇒ Research also is needed on the causes and consequences of driver turnover, as well as on how to provide incentives to drivers that might discourage them from hopping from job to job.
⇒ Research should be supported to determine which tasks are most impaired by fatigue.
⇒ How can research identify behavioral or biological markers of wake state instability and inter-individual variability in vulnerability to impairment from chronic sleep loss?

Table 12. Research Needs: Assessment (continued)

⇒ Research should determine the relationship between sleep debt and both obesity and metabolic function among workers in transportation, health care, and other industries where extended work and habitually restricted sleep are common.
⇒ There is a need for experiments to determine how many hours of sleep a person needs on the weekend (or extended break period) to recover the sleep loss from the preceding work week.
⇒ What are the long term consequences for people who do not get restorative sleep during their extended break periods? Do they ever recover from their sleep debt or do they just add to it? If so, what are the cumulative consequences for both health and safety?
⇒ Longitudinal studies of sleep and health are needed. Research on sleep mechanisms is needed to determine how sleep loss impairs metabolism and may promote obesity, insulin resistance, and hypertension so that rational counter-measures can be developed.
⇒ Research is needed to determine whether shift work is associated with depression, cancer, or sickness absence. Is this due to the lack of slow wave sleep and therefore due to lower quality sleep at odd hours?
⇒ Research must be performed to understand the impact of elevated levels of cortisol and thyroid stimulating hormone in night shift workers, which includes truck drivers. What are the safety and, perhaps most importantly, the health implications?
⇒ Research is needed to construct validated and sensitive measures of fatigue to demonstrate effective fatigue management.
⇒ Does working long or irregular hours have other harmful effects on health in addition to those related to partial sleep deprivation?
⇒ Future research should include work-based fatigue studies, physiological and biochemical monitoring studies at work and rest, assessment of the impact of rest periods on vigilance and fatigue, studies of unwinding, development of methods to reduce maladaptive coping strategies, and studies of how changes in the work environment affect hypertension.
⇒ Research should determine the extent to which research findings concerning job stress among long-distance bus drivers and urban transit workers are applicable to truck drivers.

Table 12. Research Needs: Assessment (continued)

⇒ Further research could determine if employment as a truck driver leads to increased smoking, poor diet, lack of exercise, and an increased incidence of multiple sexual partners.
⇒ Team operations require a special research focus and needs assessment. Very little is known about the health effects of team driving as a special case. While policy-makers have little information on the safety impacts, they have virtually no information on team driver health and mortality.

Only after careful data analysis, understanding of the economic forces at work in the industry or productive process, and assessment of the problem, is it useful to begin to undertake interventions. While we are all in a hurry to find some solutions to the health problems facing truck drivers, we should have this foundation before talking about interventions. Intervention before assessment of the health problems embedded in the productive process may produce interventions that do not work and might even do more harm than good.

Table 13. Research Needs: Interventions and Countermeasures

⇒ Research could explore ways of organizing the provision of medical care so that truck drivers have access to health services, including continuity of care needed for effective diagnosis and treatment.
⇒ How serious is the effect of irregular and unpredictable scheduling? How can policy-makers reduce the irregularity and unpredictability of scheduling without introducing potentially devastating costs into the logistics chain?
⇒ Will some people naturally tolerate the scheduling problems that are traditional in trucking and transportation? What personal characteristics explain differences in drivers' ability to deal with night work or irregular shifts?
⇒ What time or other intervention might help truck drivers better recover from sleep loss without harming their health?
⇒ How can drivers and similar workers better unwind after work? How can unwinding be accomplished without the use of self-medication? How can drivers be socially reintegrated after a stressful day of driving?
⇒ What are the long-term health implications of addressing drivers' inability to adjust to irregular shift schedules by using pharmacologic or other approaches that force or trick the body to conform with commercial scheduling demands?

Table 13. Research Needs: Interventions and Countermeasures (continued)

⇒ Should additional public or private highway rest areas be created to provide drivers with a safe place to stop and sleep?
⇒ How do we design and implement successful remediation programs in support of healthier, safer truck drivers?
⇒ Can research validate the proposed regulatory emphasis on restorative sleep rather than working time?
⇒ Are self-employed drivers in the U.S., like those in Australia, more likely to have safety problems? Are self-employed drivers in the U.S. more likely to have nonfatal or fatal crashes? Does unionization in Australia and Canada improve the safety records of owner-drivers
⇒ How effective are the EU's second-generation tachographs in improving truck driver occupational safety and health?
⇒ Using empirical research techniques, can researchers asses the impact of HOS regulations on both productivity and production?
⇒ What changes can be made in freight yards to reduce the high rate of permanent disability among yard workers?
⇒ Can changes in cab and sleeper berth design reduce noise exposure?
⇒ Can changes in seat or sleeper berth design reduce exposure to vibration?
⇒ Can more intensive training and supervision reduce the injury rate among inexperienced drivers, dockworkers, and yard workers? How can work be redesigned to reduce the negative health and safety consequences of current work methods?
⇒ How can trucking companies establish a "corporate culture" that values safety?
⇒ Research should be undertaken to evaluate potential countermeasures, such as wellness-counseling and fitness program for professional drivers.

Finally, after all these steps are complete we can turn to education. Education will be more successful, and the acceptance of research results more likely, if all the players in the industry are part of the original process of data collection, economic analysis, assessment, and intervention. If research does not involve these critically interested parties, and if research does not integrate these elements, research findings and developments may not be well understood or may be ignored.

Table 14. Research Needs: Dissemination and Education

⇒ How can the research community get the attention of drivers themselves? How can drivers better communicate their needs to researchers? Can scholars design collaborative research that engages the trucking community so that they believe their full range of interests have been considered?
⇒ Can a small business training program affect the safety records of owner-operators and small motor carriers? Are owner-operators and small carriers that understand their costs less likely to face intense financial pressures that lead them to drive when they should be sleeping?

It would be helpful to policy makers to have a coherent strategy for addressing research needs related to truck driver occupational safety and health. Ideally, researchers could develop proposed projects that are relevant to the most pressing concerns of drivers, trucking companies, shippers, government regulators, and other interested parties; and then, a group of researchers with a wide range of expertise, working with industry, labor, and government representatives, could rank the projects in order of priority. The April 2003 conference was too short to achieve that objective. Our hope, however, is that there will be an ongoing series of conferences, collaborative interactions, and research projects to complete this task – especially ones involving cooperation between researchers and the industry. The April 2003 conference, however, was an important step in focusing attention on the problem. To quote Winston Churchill, "This is not the end. It is not even the beginning of the end. But it is, perhaps, the end of the beginning."[156]

A central theme of this conference report and literature review has been the need for more research on truck driver occupational safety and health. The July 2004 ruling by the U.S. Circuit Court of Appeals, striking down the 2003 amendment to HOS regulations on grounds that FMCSA had not given sufficient attention to the health of drivers,[157] underscores the urgency of research on not only the prevention of highway crashes, but also the prevention of job-related diseases, injuries from loading or unloading trucks, and other occupational safety and health problems faced by truck drivers. The ultimate goal, however, is to achieve improvements in the lives of truck drivers and their families. It is not easy to develop effective remedies for truck driver occupational safety and health problems that avoid imposing undue operational or economic burdens on drivers, carriers, or shippers. This is our challenge.

References

1. 49 CFR Parts 385, 390 and 395: Hours of Service of Drivers; Final Rule. Federal Register, Vol. 70, No. 164, August 25 2005, p. 49978-50073 at p. 49981. Available online on October 27, 2006 at *http://a257.g.akamaitech.net/7/257/2422/01jan20051800/edocket.access.gpo.gov/2005/pdf/05-16498.pdf.*

2. U.S. Department of Labor: Bureau of Labor Statistics. Census of Fatal Occupational Injuries, Table A-5: Fatal occupational injuries by occupation and event or exposure, All United States, 2004. Available online on July 1, 2006 at *http://www.bls.gov/iif/oshwc/cfoi/cftb0200.pdf.*

3. U.S. Department of Labor: Bureau of Labor Statistics. Lost-Worktime Injuries and Illnesses: Characteristics and Resulting Time Away from Work, 2004, Table 4: Number of nonfatal occupational injuries and illnesses involving days away from work by selected worker occupation and major industry sector, 2004. Available online on October 27, 2006 at *http://www.bls.gov/iif/oshwc/osh/case/osnr0024.pdf.*

4. U.S. Department of Labor: Bureau of Labor Statistics. May 2004 National Occupational Employment and Wage Estimates, United States. Available online on October 11, 2006 at *http://www.bls.gov/oes/oes_2004_m.htm.*

5. U.S. Department of Labor: Bureau of Labor Statistics. National Census of Fatal Occupational Injuries in 2004. Available online October 27, 2006 at *http://www.bls.gov/iif/oshwc/cfoi/cfnr0011.pdf*.

6. Panzar JC, Willig RD. Economies of Scope. American Economic Review, Vol. 71, No. 2, May 1981, pp. 268-272.

7. Jara-Diaz SR, Basso LJ. Transport Cost Functions, Network Expansion, and Economies of Scope. Transportation Research, Part E Logistics and Transportation Review, Vol. 39, No. 4, 2003, pp. 271-288.

8. Keaton MH. Are There Economies of Traffic Density in the Less-than-Truckload Motor Carrier Industry? An Operations Planning Analysis. Proceedings of the Transportation Research Forum, Vol. 27A, No. 5, 1993, pp. 343-358.

9. Stigler GJ. The Economies of Scale. Journal of Law and Economics, Vol. 1, No. 1, October 1958, pp. 54-71.

10. Giordano JN. Returns to Scale and Market Concentration among the Largest Survivors of Deregulation in the US Trucking Industry. Applied Economics, Vol. 29, No. 1, 1997, pp. 101-110.

11. Dean F. The Ground War at FedEx: Drivers are suing to protest their status as contractors -- and gaining traction. Business Week, No. 3961, 2005, p. 42.

12. Roadway Package System, Inc., 326 NLRB No. 72, 159 LRRM 1153 (1998) and Dial-A-Mattress Operating Corporation, 326 NLRB No. 75, 159 LRRM 1166 (1998), following the common-law agency standard set by Supreme Court in NLRB v. United Insurance Co., 390 U.S. 254, 257 (1968).

13. Belzer MH. Sweatshops on Wheels: Winners and Losers in Trucking Deregulation (New York: Oxford University Press, 2000).

14. Belzer MH. Collective Bargaining in the Trucking Industry: Do the Teamsters Still Count? Industrial and Labor Relations Review, Vol. 48, No. 4; July 1995, pp. 636-655.

15. Belzer MH. Trucking: Collective Bargaining Takes a Rocky Road , in Paul F. Clark, John T. Delaney, and Ann C. Frost (eds.), Collective Bargaining in the Private Sector (Industrial Relations Research Association Series; Champaign, IL: Industrial Relations Research Association, 2002), 311-42.

16. Belzer MH. Sweatshops on Wheels, op cit. n 13, p. 47.

17. Belman DL, Monaco KA. The Effects of Deregulation, De-Unionization, Technology, and Human Capital on the Work and Work Lives of Truck Drivers. Industrial and Labor Relations Review, Vol. 54, No. 2A, March 2001, pp. 502-525

18. Belzer MH. Hours of Service Impact Assessment. University of Michigan Transportation Research Institute, March 5, 1999. With Kenneth Campbell, Stephen Burks, Kristen Monaco, George Fulton, Donald Grimes, Daniel Lass, and Dale Ballou. Research performed for Federal Highway Authority, Office of Motor Carriers and Highway Safety.

19. Belzer MH. Paying the Toll: Economic Deregulation of the Trucking Industry. Washington, D.C.: Economic Policy Institute. 1994. Briefing and Working Papers Series.

20. Budd JW. Employment with a Human Face (Ithaca, NY: Cornell University Press, 2004).

21. Belzer MH, Rodriguez DA, Sedo SA. Paying for Safety: An Economic Analysis of the Effect of Compensation on Truck Driver Safety. Washington, DC: United States Department of Transportation, Federal Motor Carrier Safety Administration; 2002. Available on line on August 18, 2006 at http://ai.volpe.dot.gov/CarrierResearchResults/CarrierResearchContent.stm#car2.

22. Akerlof GA, Dickens WT. The Economic Consequences of Cognitive Dissonance. American Economic Review, Vol. 72, No. 3, June 1982, pp. 307-319.

23. Quinlan, M. Report of Inquiry into Safety in the Long Haul Trucking Industry. Motor Accidents Authority of New South Wales, Sydney. Available on line December 12, 2006 at *http://www.maa.nsw.gov.au/default.aspx?MenuID=189#171* and included on the accompanying CD-ROM.

24. Directive 2002/15/EC of the European Parliament and of the Council of 11 March 2002 on the organisation of the working time of persons performing mobile road transport activities. Official Journal of the European Communities, L80, Vol 45, 35-39. Available online October 4, 2006 at: http://europa.eu.int/eur-lex/pri/en/oj/dat/2002/l_080/l_08020020323en00350039.pdf

25. 49 CFR Parts 385, 390, and 395: Hours of Service of Drivers; Driver Rest and Sleep for Safe Operations; Final Rule. Federal Register, Vol. 68, No. 81, April 28 2003, pp. 22456-22517 at p. 22495.

26. Thompson K. Hours of Service: Understanding the Changes. Hours of Service Productivity Summit. Georgia Institute of Technology. Atlanta, GA. October 30, 2003. Accessed on July 13, 2006, at: http://tli.isye.gatech.edu/HOS/archive/TLI_Schneider-Kirk_Thompson.pdf.

27. Gifford T. Impact of New HOS on Transportation Costs. Hours of Service Productivity Summit. Georgia Institute of Technology. Atlanta, GA. October 30, 2003. Accessed on July 13, 2006, at: http://tli.isye.gatech.edu/HOS/archive/TLI_Schneider-Ted_Gifford.pdf.

28. European Transport Safety Council. The Role of Driver Fatigue in Commercial Road Transport Crashes. Brussels, 2001. Available online on October 27, 2006 at *http://www.etsc.be/documents/drivfatigue.pdf*.

29. Cambois E, Robine JM, Hayward, MD. Social Inequalities in Disability-Free Life Expectancy in the French Male Population, 1980-1991. Demography, Vol. 38, No. 4, November 2001, pp. 513-524.

30. Pappas G, Queen S, Wilbur Hadden, and Gail Fisher. The Increasing Disparity in Mortality between Socioeconomic Groups in the United States, 1960 and 1986. New England Journal of Medicine, Vol. 329, No. 2, July 8, 1993, pp. 103-109.

31. Adler NE, Boyce WT, Chesney MA, et al. Socioeconomic Inequalities in Health: No Easy Solution. JAMA, Vol 269, No. 24, June 23, 1993, pp. 3140-3145.

32. Lantz PM, House JS, Lepkowski JM, et al. Socioeconomic Factors, Health Behaviors, and Mortality: Results from a Nationally Representative Prospective Study of U.S. Adults. JAMA, Vol. 279, No. 21, June 3, 1998, pp. 1703-1708.

33. North FM, Syme LS, Feeney A, et al. Psychosocial Work Environment and Sickness Absence among British Civil Servants: The Whitehall II Study. American Journal of Public Health, Vol. 86, No 3, March 1996, pp. 332-340.

34. Marmot MG, Bosma H, Hemingway H, et al. Contribution of Job Control and Other Risk Factors to Social Variations in Coronary Heart Disease Incidence. The Lancet, Vol. 350, No. 9073, July 26, 1997, pp. 235-239.

35. Fuhrer R, Shipley MJ, Chastang JF, et al. Socioeconomic Position, Health, and Possible Explanations: A Tale of Two Cohorts. American Journal of Public Health, Vol. 92, No 8, August 2002, pp. 1290-1294.

36. Park RM, Bailer AJ, Stayner LT, Halperin W, Gilbert SJ. An Alternate Characterization of Hazard in Occupational Epidemiology: Years of Life Lost Per Years Worked. American Journal of Industrial Medicine, Vol. 42, No. 1, July 2002, pp. 1-10.

37. Saltzman GM, Belzer MH. The Case for Strengthened Motor Carrier Hours of Service Regulations. Transportation Journal, Vol. 41, No. 4, Summer 2002, pp. 51-71 at p. 62.

38. Aronson KJ, Howe GR, Carpenter M, and Fair ME. Surveillance of Potential Associations between Occupations and Causes of Death in Canada, 1965-91. Occupational and Environmental Medicine, Vol. 56, No. 4, April 1999, pp. 265-269.

39. Hansen ES. A Follow-Up Study on the Mortality of Truck Drivers. American Journal of Industrial Medicine, Vol. 23, No 5, May 1993, pp. 811-821.

40. Hannerz H, Tuchsen F. Hospital Admissions among Male Drivers in Denmark. Occupational and Environmental Medicine, Vol. 58, No. 4, April 1, 2001, pp. 253-260.

41. Hakkola M, Honkasalo ML, Pulkkinen P. Changes in Neuropsychological Symptoms and Moods among Tanker Drivers Exposed to Gasoline during a Work Week. Occupational Medicine (London), Vol. 47, No. 6, August 1997, pp. 344-348.

42. Harrington JM. Health Effects of Shift Work and Extended Hours of Work. Occupational and Environmental Medicine, Vol. 58, No. 1, January 2001, pp. 68-72 at p. 71.

43. Hoek G, Brunekreef B, et al. Association between Mortality and Indicators of Traffic-Related Air Pollution in the Netherlands: A Cohort Study. The Lancet, Vol. 360, No. 9341, October 19, 2002, pp. 1203-1209.

44. Clark N, Dropkin J, Kaplan L. Ready Mixed Concrete Truck Drivers: Work-Related Hazards and Recommendations for Controls. unpublished report, Mount Sinai-Irving J. Selikoff Center for Occupational and Environmental Medicine, September 2001, p. 5 (included on the accompanying CD-ROM and at *http://www.ilir.umich.edu/TIBP/truckdriverOSH*).

45. Ibid., p. 6.

46. Ibid., p. 7.

47. Van den Heever DJ, Roets FJ. Noise Exposure of Truck Drivers: A Comparative Study. American Industrial Hygiene Association Journal, Vol. 57, No. 6, June 1996, pp. 564-566.

48. Sheshagiri B. Occupational Noise Exposure of Operators of Heavy Trucks. American Industrial Hygiene Association Journal, Vol. 59, No. 3, March 1998, pp. 205-213.

49. Palmer KT, Griffin MJ, Bendall H, et al. Prevalence and Pattern of Occupational Exposure to Whole Body Vibration in Britain: Findings from a National Survey. Occupational and Environmental Medicine, Vol. 57, No. 4, April 2000, pp. 229-236.

50. Pope MH, Magnusson M, Wilder DG. Kappa Delta Award. Low Back Pain and Whole Body Vibration. Clinical Orthopaedics and Related Research, No. 354, September 1998, pp. 241-248.

51. Pietri F, Leclerk A, Boitel L, et al. Low-Back Pain in Commercial Travelers. Scandinavian Journal of Work, Environment, and Health, Vol. 18, No. 1, February 1992, pp. 52-58.

52. Jensen MV, Tuchsen F, Orhede E. Prolapsed Cervical Intervertebral Disc in Male Professional Drivers in Denmark, 1981-1990. Spine, Vol. 21, No. 20, October 15, 1996, pp. 2352-2355.

53. Rodriguez DA, Rocha M, Khattak AJ, Belzer MH. Effects of Truck Driver Wages and Working Conditions on Highway Safety: Case Study. Transportation Research Record, Freight Policy, Economics, and Logistics; Truck Transportation, No. 1833, 2003, pp. 95-102.

54. Rodriguez DA, Targa F, Belzer MH. Pay Incentives and Truck Driver Safety: A Case Study. Industrial and Labor Relations Review, Vol. 59, No. 2, 2006, pp. 205-225.

55. Hamelin P. Professional Drivers' Working Time as a Factor of Flexibility and Competitiveness in Road Haulage. TUTB Newsletter, Vol. 15-16, 2001, pp. 39-47 at 43. Accessed on October 11, 2006 at: *http://hesa.etui-rehs.org/uk/newsletter/files/Newsletter-15.pdf*.

56. Greenhouse S. Bush to Open Country to Mexican Truckers. The New York Times, February 7, 2001, p. A12.

57. Public Citizen v. Department of Transportation, 316 F.3d 1002 (9th Circuit, 2003).

58. Department of Transportation v. Public Citizen, 541 U.S. 752 (2004).

59. Harmon D. For Most Mexican Truckers, Access to US a Waiting Game. Christian Science Monitor, February 17, 2005, p. 7.

60. Folkard S, Tucker P. Shift Work, Safety, and Productivity. Occupational Medicine (London), Vol. 53, No. 2, March 2003, pp. 95-101.

61. Tucker P, Folkard S, McDonald I. Rest Breaks and Accident Risk. The Lancet, Vol. 361, No. 9358, February 22, 2003, p. 680.

62. Folkard S, Lombardi DA. Toward a 'Risk Index' to Assess Work Schedules. Chronobiology International, Vol. 21, No. 6, 2004, pp. 1063-1072.

63. Dinges DF, Pack F, Williams K, et al. Cumulative Sleepiness, Mood Disturbance, and Psychomotor Vigilance Performance Decrements during a Week of Sleep Restricted to 4-5 Hours Per Night, Sleep, Vol. 20, No. 4, April 1997, pp. 267-277.

64. Folkard S, Akerstedt T. Trends in the Risk of Accidents and Injuries and Their Implications for Models of Fatigue and Performance. Aviation, Space and Environmental Medicine, Vol. 75, No. 3, Section II, March 2004, pp. A161-A167.

65. Freund DM. An Annotated Literature Review Relating to Proposed Revisions to the Hours-of-Service Regulation for Commercial Motor Vehicle Drivers. Office of Motor Carrier Safety, U.S. Department of Transportation, November 1999, DOT-MC-99-129.

66. Hamelin P. Lorry Driver's Time Habits in Work and Their Involvement in Traffic Accidents. Ergonomics, Vol. 30. No. 9, September 1987, pp. 1323-1333.

67. Akerstedt T, Czeisler CA, Dinges DF, Horne JA. Accidents and Sleepiness: A Consensus Statement from the International Conference on Work Hours, Sleepiness and Accidents, Stockholm, 8-10 September 1994. Journal of Sleep Research, Vol. 3, 1994, p. 195.

68. Stoohs RA, Guilleminault C, Itoi A, Dement WC. Traffic Accidents in Commercial Long-Haul Truck Drivers: The Influence of Sleep-Disordered Breathing and Obesity. Sleep, Vol. 17, No. 7, October 1994, pp. 619-623.

69. Folkard S. Black Times: Temporal Determinants of Transport Safety. Accident Analysis and Prevention, Vol. 29, No. 4, July 1997, pp. 417-430.

70. Arnold PK, Hartley LR, Corry A, Hochstadt D, Penna F, Feyer AM. Hours of Work, and Perceptions of Fatigue among Truck Drivers. Accident Analysis and Prevention, Vol. 29, No. 4, 1997, pp. 471-477.

71. McCartt AT, Rohrbaugh JW, Hammer MC, Fuller SZ. Factors Associated with Falling Asleep at the Wheel among Long-Distance Truck Drivers. Accident Analysis and Prevention, Vol. 32, No. 4, July 2000, pp. 493-504.

72. Lyznicki JM, Doege TC, David RM, Williams MA. Sleepiness, Driving, and Motor Vehicle Crashes. JAMA, Vol. 279, No. 23, June 17, 1998, pp. 1908-1913.

73. Dement WC. The Perils of Drowsy Driving. New England Journal of Medicine, Vol. 337, No. 11, September 11, 1997, pp. 783-785.

74. Barach P, David GB, Richter E. The Sleep of Long-Haul Truck Drivers (Correspondence). New England Journal of Medicine, Vol. 338, No. 6, February 5, 1998, pp. 389-391.

75. Summala H, Mikkola T. Fatal Accidents among Car and Truck Drivers: Effects of Fatigue, Age, and Alcohol Consumption. Human Factors, Vol. 36, No. 2, June 1994, pp. 315-326.

76. Hakkanen H, Summala H. Sleepiness at Work among Commercial Truck Drivers. Sleep, Vol. 23, No. 1, February 1, 2000, pp. 49-57.

77. Philip P, Vervialle F, Le Breton P, et al. Fatigue, Alcohol, and Serious Road Crashes in France: Factorial Study of National Data. British Medical Journal, Vol. 322, No. 7290, April 7, 2001, pp. 829-830.

78. Feyer AM. Fatigue: Time to Recognize and Deal with an Old Problem: It's Time to Stop Treating Lack of Sleep As a Badge of Honor. British Medical Journal, Vol. 322, No. 7290, April 7, 2001, pp. 808-809.

79. Horne JA, Reyner LA. Sleep Related Vehicle Accidents. British Medical Journal, Vol. 310, No. 6979, March 4, 1995, pp. 565-567.

80. Horne JA, Reyner LA. Vehicle Accidents Related to Sleep: A Review. Occupational and Environmental Medicine, Vol. 56, No. 5, May 1999, pp. 289-294.

81. Dawson D, Reid KJ. Fatigue, Alcohol and Performance Impairment. Nature, Vol. 388, No 6639, July 17, 1997, p. 235.

82. Lamond N, Dorrian J, Roach GD, et al. Adaptation of Performance during a Week of Simulated Night Work. Ergonomics, Vol. 47, No. 2, February 2004, pp. 154-165..

83. Dawson D, Lamond N, Donkin K, Reid KJ. Quantitative Similarity between the Cognitive Psychomotor Performance Decrement Associated with Sustained Wakefulness and Alcohol Intoxication. in Laurence Hartley (ed.), Managing Fatigue in Transportation (Oxford: Elsevier Science, 1998).

84. Dawson D, Reid KJ. Fatigue Management in Aviation: Similarities between the Effects of Fatigue and Alcohol on Performance Impairment. in Brent J. Hayward and Andrew R. Lowe (eds.), Aviation Resource Management (Aldershot, England: Ashgate Publishing, 2000).

85. Lamond N, Dawson D. Quantifying the Performance Impairment Associated with Fatigue. Journal of Sleep Research, Vol. 8, No. 4, December 1999, pp. 255-262.

86. Roach GD, Dorrian J, Fletcher A, Dawson D. Comparing the Effects of Fatigue and Alcohol Intoxication on Locomotive Engineers' Performance in a Rail Simulator. Journal of Human Ergology, Vol. 30, Nos. 1-2, 2001, pp. 121-126.

87. Williamson AM, Feyer AM. Moderate Sleep Deprivation Produces Impairments in Cognitive and Motor Performance Equivalent to Legally Prescribed Levels of Alcohol Intoxication. Occupational and Environmental Medicine, Vol. 57, No. 10, October 1, 2000, pp. 649-655.

88. Williamson AM, Feyer AM, Mattick RP, et al. Developing Measures of Fatigue Using an Alcohol Comparison to Validate the Effects of Fatigue on Performance. Accident Analysis and Prevention, Vol. 33, No. 3, May 2001, pp. 313-326.

89. Arnedt JT, Wilde GJS, Munt PW, MacLean AW. How Do Prolonged Wakefulness and Alcohol Compare in the Decrements They Produce on a Simulated Driving Task? Accident Analysis and Prevention, Vol. 33, No. 3, May 2001, pp. 337-344.

90. Van Dongen HPA, Maislin G, Mullington JM, Dinges DF. The Cumulative Cost of Additional Wakefulness: Dose-Response Effects on Neurobehavioral Functions and Sleep Physiology from Chronic Sleep Restriction and Total Sleep Deprivation. Sleep, Vol. 26, No. 2, March 15, 2003, pp. 117–126.

91. Belenky G, Wesensten NJ, Thorne DR, et al. Patterns of Performance Degradation and Restoration during Sleep Restriction and Subsequent Recovery: A Sleep Dose-Response Study. Journal of Sleep Research, Vol. 12, No. 1, March 2003, pp. 1–12.

92. Doran SM, Van Dongen HPA, Dinges DF. Sustained Attention Performance during Sleep Deprivation: Evidence of State Instability. Archives Italiennes de Biologie, Vol. 139, No. 3, April 2001, pp. 253–267.

93. Aeschbach D, C. Cajochen, H. Landolt, and A.A. Borbély. Homeostatic Sleep Regulation in Habitual Short Sleepers and Long Sleepers. American Journal of Physiology, Vol. 270, No. 1, Pt. 2, January 1996, pp. R41–R53.

94. Van Dongen HPA, Rogers NL, Dinges DF. Sleep Debt: Theoretical and Empirical Issues. Sleep and Biological Rhythms, Vol. 1, 2003, pp. 5–13.

95. Buxton OM, Spiegel K, Van Cauter E. Modulation of Endocrine Function and Metabolism by Sleep and Sleep Loss. In: T. Lee-Chiong, M. Carskadon, M. Sateia, vol. eds., Sleep Medicine 2002; 59-69.

96. Spiegel K, Leproult R, Van Cauter E. Impact of Sleep Debt on Metabolic and Endocrine Function. The Lancet, Vol. 354, No. 9188, October 23, 1999, pp. 1435-1439.

97. Spiegel K, Leproult R, Colecchia EF, et al. Adaptation of the 24-h Growth Hormone Profile to a State of Sleep Debt. American Journal of Physiology - Regulatory Integrative & Comparative Physiology, Vol. 279, No. 3, September 2000, pp. R874-883.

98. Spiegel K, Sheridan JF, Van Cauter E. Effect of Sleep Deprivation on Response to Immunization. JAMA, Vol. 288, No. 12, September 25, 2002, pp. 1471-1472.

99. Stoohs RA, Bingham LA, Itoi A, Guilleminault C, Dement WC. Sleep and Sleep-Disordered Breathing in Commercial Long-Haul Truck Drivers. Chest, Vol 107, No. 7, May 1995, pp. 1275-1282.

100. Stoohs RA, Dement WC. Snoring and Sleep-Related Breathing Abnormality During Partial Sleep Deprivation (Correspondence). New England Journal of Medicine, Vol. 328, No. 17, April 29, 1993, p. 1279.

101. Dingus TA, Hardee HL, Wierwille WW. Development of Models for On-Board Detection of Driver Impairment. Accident Analysis and Prevention, Vol. 19, No. 4, August 1987, pp. 271-283.

102. Taylor PJ, Pocock SJ. Mortality of Shift and Day Workers 1956-1968. British Journal of Industrial Medicine, Vol. 29, No. 2, April 1972, pp. 201-207.

103. Williamson AM, Feyer AM, Friswell R. The Impact of Work Practices on Fatigue in Long Distance Truck Drivers. Accident Analysis and Prevention, Vol. 28, No. 6, November 1996, pp. 709-719.

104. Feyer AM, Williamson A, Friswell R. Balancing Work and Rest to Combat Driver Fatigue: An Investigation of Two-Up Driving in Australia. Accident Analysis and Prevention, Vol. 29, No. 4, July 1997, pp. 541-553.

105. Williamson and Feyer, op cit. n. 87, pp. 653-654.

106. Mitler MM, Miller JC, Lipsitz JJ, et al. The Sleep of Long-Haul Truck Drivers. New England Journal of Medicine, Vol. 337, No. 11, September 11, 1997, pp. 755-761.

107. Hanowski RJ, Wierwille WW, Dingus TA. An On-Road Study to Investigate Fatigue in Local/Short Haul Trucking. Accident Analysis and Prevention, Vol. 35, No. 2, 2003, pp. 153-160.

108. Brown ID. Driver Fatigue. Human Factors, Vol. 36, No. 2., 1994, pp. 298-314 at p. 302.

109. Beilock R. Schedule-Induced Hours-of-Service and Speed Limit Violations among Tractor-Trailer Drivers. Accident Analysis and Prevention, Vol. 27, No. 1, 1995, pp. 33-42.

110. Torsvall L, Akerstedt T. Disturbed Sleep While Being on Call: An EEG Study of Ships' Engineers. Sleep, Vol. 11, No. 1, February 1988, pp. 35-38.

111. Akerstedt T, Knutsson A, Westerholm P, et al. Sleep Disturbances, Work Stress and Work Hours: A Cross-Sectional Study. Journal of Psychosomatic Research, Vol. 53, No. 3, 2002, pp. 741-748.

112. Cziesler CA, Moore-Ede MC, Coleman RM. Rotating Shift Work Schedules That Disrupt Sleep Are Improved by Applying Circadian Principles. Science, Vol. 217, No. 4558, July 30, 1982, pp. 460-463.

113. Monk TH. Advantages and Disadvantages of Rapidly Rotating Shift Schedules—A Circadian Viewpoint. Human Factors, Vol. 28, No. 5, October 1986, pp. 553-557.

114. Knauth P. The Design of Shift Systems. Ergonomics, Vol. 36, Nos. 1-3, 1993, pp. 15-28.

115. Akerstedt T. Shift Work and Disturbed Sleep/Wakefulness. Occupational Medicine (London), Vol. 53, No. 2, March 2003, pp. 89-94.

116. Muller v. Oregon, 208 US 412 (1908).

117. Spelten E, Barton J, Folkard S. Have We Underestimated Shiftworkers' Problems? Evidence from a 'Reminiscence' Study. Ergonomics, Vol. 36, Nos. 1-3, January-March 1993, pp. 307-312.

118. Ibid. at p. 307.

119. Leproult R, Copinschi G, Buxton O, Van Cauter E. Sleep Loss Results in Elevation of Cortisol Levels the Next Evening. Sleep, Vol. 20, No. 10, October 1997, pp. 865-870.

120. Tuchsen F. Stroke Morbidity in Professional Drivers in Denmark 1981-1990. International Journal of Epidemiology, Vol. 26, No. 5, October 1997, pp. 989-994.

121. Tochikubo O, Ikeda A, Miyajima E, Ishii M. Effects of Insufficient Sleep on Blood Pressure Monitored by a New Multibiomedical Recorder. Hypertension, Vol. 27, No 6, June 1996, pp. 1318-1324.

122. Irwin M, McClintick J, Costlow C, et al. Partial Night Sleep Deprivation Reduces Natural Killer and Cellular Immune Responses in Humans. FASEB Journal, Vol. 10, No. 5, April 1996, pp. 643-653 at p. 643.

123. Irwin M, Thompson J, Miller C, et al. Effects of Sleep and Sleep Deprivation on Catecholamine and Interleukin-2 Levels in Humans: Clinical Implications. The

Journal of Endocrinology and Metabolism, Vol. 84, No. 6, June 1999, pp. 1979-1985.

124. Tenkanen L, Sjoblom T, Harma M. Joint Effect of Shift Work and Adverse Life-Style Factors on the Risk of Coronary Heart Disease, Scandinavian Journal of Work, Environment, and Health, Vol. 24, No 5, October 1998, pp. 351-357.

125. Michie S, Cockcroft A. Overwork Can Kill: Especially If Combined with High Demand, Low Control, and Poor Social Support. British Medical Journal, Vol. 312, No. 7036, April 13, 1996, pp. 921-922.

126. Spurgeon A, Harrington JM, Cooper CL. Health and Safety Problems Associated with Long Working Hours: A Review of the Current Position. Occupational and Environmental Medicine, Vol. 54, No. 6, June 1997, pp. 367-375.

127. Harrington JM. Health Effects of Shift Work and Extended Hours of Work. Occupational and Environmental Medicine, Vol. 58, No. 1, January 2001, pp. 68-73.

128. Sparks K, Cooper CL, Fried Y, Shirom A. The Effects of Hours of Work on Health: A Meta-Analytic Review. Journal of Occupational and Organizational Psychology, Vol. 70, No. 4, December 1997, pp. 391-408.

129. Nylen L, Voss M, Floderus B. Mortality among Women and Men Relative to Unemployment, Part-Time Work, Overtime Work, and Extra Work: A Study Based on Data from the Swedish Twin Registry. Occupational and Environmental Medicine, Vol. 58, No. 1, January 2001, pp. 52-57.

130. Knutsson A. Health Disorders of Shift Workers. Occupational Medicine (London), Vol. 53, No. 2, March 2003, pp. 103-108.

131. Stoynev AG, Minkova NK. Circadian Rhythms of Arterial Pressure, Heart Rate and Oral Temperature in Truck Drivers. Occupational Medicine (London), Vol. 47, No. 3, April 1997, pp. 151-154.

132. Tuchsen F. Working Hours and Ischaemic Heart Disease in Danish Men: A 4-Year Cohort Study of Hospitalization. International Journal of Epidemiology, Vol. 22, No. 2, April 1993, pp. 215-221.

133. Emdad R, Belkic K, Theorell T, et al. Work Environment, Neurophysiologic and Psychophysiologic Models among Professional Drivers with and without Cardiovascular Disease: Seeking an Integrative Neurocardiologic Approach. Stress Medicine, Vol. 13, No. 1, January 1997, pp. 7-21.

134. Emdad R, Belkic K, Theorell T, Cizinsky S. What Prevents Professional Drivers from Following Physicians' Cardiologic Advice?. Psychotherapy & Psychosomatics, Vol. 67, Nos. 4-5, July-October 1998, pp. 226-240.

135. Rosenstock SJ, Anderson LP, Rosenstock CV, et al. Socioeconomic Factors in Helicobacter pylori Infection among Danish Adults. American Journal of Public Health, Vol. 86, No. 11, November 1996, pp. 1539-1544.

136. Kuiper JI, Van der Beek AJ, Meijman TF. Psychosomatic Complaints and Unwinding of Sympathoadrenal Activation after Work. Stress Medicine, Vol. 14, No. 1, January 1998, pp. 7-12.

137. Van der Beek AJ, Meijman TF, Frings-Dresen MHW, et al. Lorry Drivers' Work Stress Evaluated by Catecholamines Excreted in Urine. Occupational and Environmental Medicine, Vol. 52, No. 7, July 1995, pp. 464-469.

138. Orris P, Hartman DE, Strauss P, et al. Stress among Package Truck Drivers. American Journal of Industrial Medicine, Vol. 31, No. 2, February 1997, pp. 202-210.

139. Raggatt PT, Morrissey SA. A Field Study of Stress and Fatigue in Long-Distance Bus Drivers. Behavioral Medicine, Vol. 23, No. 3, Fall 1997, pp. 122-129.

140. Sluiter JK, Van der Beek AJ, Frings-Dresen MH. Work Stress and Recovery Measured by Urinary Catecholamines and Cortisol Excretion in Long Distance Coach Drivers. Occupational and Environmental Medicine, Vol. 55, No. 6, June 1998, pp. 407-413.

141. Costa G. Shift Work and Occupational Medicine: An Overview. Occupational Medicine (London), Vol. 53, No 2, March 2003, pp. 83-88 at p. 85.

142. Hedberg GE, Wikstrom-Frisen L, Janlert U. Comparison between Two Programmes for Reducing the Levels of Risk Indicators of Heart Diseases among Male Professional Drivers. Occupational and Environmental Medicine, Vol. 55, No. 8, August 1998, pp. 554-561.

143. Solomon AJ, Doucette JT, Garland E, McGinn T. Health Care and the Long Haul: Long Distance Truck Drivers—A Medically Underserved Population. American Journal of Industrial Medicine, Vol. 46, No. 5, November 2004, pp. 463-471.

144. Ibid. at p. 467.

145. Rosa RR, Bonnet MH, Bootzin RR, et al. Intervention Factors for Promoting Adjustment to Nightwork and Shiftwork. Occupational Medicine, Vol. 5, No. 2, April-June 1990, pp. 391-415.

146. Knauth P, Hornberger S. Preventive and Compensatory Measures for Shift Workers. Occupational Medicine (London), Vol. 53, No. 2, March 2003, pp. 109-116.

147. Horowitz TS, Cade BE, Wolfe JM, Czeiser CA. Efficacy of Bright Light and Sleep/Darkness Scheduling in Alleviating Circadian Maladaptation to Night Work. American Journal of Physiology – Endocrinology & Metabolism, Vol. 281, No. 2. August 2001, pp. E384-E391.

148. Costa, op cit. n. 141, p. 86.

149. Seidel WF, Roth T, Roehrs T, Zorick F, Dement WC. Treatment of a 12-Hour Shift of Sleep Schedule with Benzodiazepines. Science, Vo. 224, No. 4654, June 15, 1984, pp. 1262-1264.

150. Arendt J, Aldhous M, English J, et al. Some Effects of Jet-Lag and Their Alleviation by Melatonin. Ergonomics, Vol. 30, No. 9, September 1987, pp. 1379-1393.

151. Van Cauter E, Turek FW. Strategies for Resetting the Human Circadian Clock. New England Journal of Medicine, Vol. 322, No. 18, May 3, 1990, pp. 1306-1308.

152. Czeisler CA, Johnson MP, Duffy JF, Brown EN, Ronda JM, Kronauer RE. Exposure to Bright Light and Darkness to Treat Physiologic Maladaptation to Night Work, New England Journal of Medicine, Vol. 322, No. 18, May 3, 1990, pp. 1253-1259.

153. U.S. Department of Transportation: Federal Highway Administration. Commercial Vehicle Driver Survey: Assessment of Parking Needs and Preferences (Washington, DC: Federal Highway Administration, U.S. Department of Transportation), March 2002, Report No. FHWA-RD-01-160

154. Kochan TA, Dyer L, Lipsky DB. The Effectiveness of Union-Management Safety and Health Committees (Kalamazoo, MI: W.E. Upjohn Institute for Employment Research, 1977).

155. U.S. Department of Transportation: Federal Motor Carrier Safety Administration. Impact of Sleeper Berth Usage on Driver Fatigue: Final Report. TechBrief Publication No. FMCSA-MCRT-02-070, August 2002. Available online October 16, 2006 at: *http://www.fmcsa.dot.gov/facts-research/research-technology/tech/Sleeper-Berth-TechBrief.pdf*.

156. Churchill W. Speech at the Lord Mayor's luncheon (London), November 10, 1942, quoted in http://www.churchill-society-london.org.uk/EndoBegn.html (available online, June 30, 2006).

157. Public Citizen v. Federal Motor Carrier Safety Administration, 374 F.3d 1209 at 1216 (D.C. Circuit, 2004).

Appendix 1: Conference Participants and Speaker Profiles

| \multicolumn{4}{c}{Participants, April 24–25, 2003, Conference on Truck Driver Occupational Safety and Health} |

Name	Title	Institution	City
Torbjorn Akerstedt	Professor, Public Health Sciences	Karolinska Institutet	Stockholm, Sweden
Toni Alterman	Chief, Illness Effects Section, now Senior Epidemiologist	Centers for Disease Control, National Institute for Occupational Safety and Health	Cincinnati, OH
Albert Alvarez	Transportation Specialist, Research Division	Federal Motor Carrier Safety Administration	Washington, DC
Michael Belzer	Associate Professor, Industrial Relations	Wayne State University	Detroit, MI
Elyce Biddle	Chief Statistician	National Institute for Occupational Safety and Health	Morgantown, WV
Stephen Burks	Assistant Professor and now Associate Professor, Economics and Management	University of Minnesota, Morris	Morris, MN
Orfeu Buxton	Instructor, Division of Sleep Medicine and Associate Neuroscientist	Harvard Medical School and Brigham & Women's Hospital	Boston, MA
Claire Caruso	Research Health Scientist	National Institute for Occupational Safety and Health	Cincinnati, OH
Guang Chen	Senior Service Fellow, Division of Safety Research	National Institute for Occupational Safety and Health	Morgantown, WV
Ralph Craft	Crash Data Analyst, Analysis Division	Federal Motor Carrier Safety Administration	Fairfax, VA

\multicolumn{4}{c}{**Participants, April 24–25, 2003, Conference on Truck Driver Occupational Safety and Health**}
Name	**Title**	**Institution**	**City**
Drew Dawson	Professor, Centre for Sleep Research	University of South Australia	Adelaide, Australia
Christopher Drake	Senior Bioscientific Staff, Sleep Disorders Research Center and now Senior Staff Scientist at the Henry Ford Hospital Sleep Center and Assistant Professor of Psychiatry and Behavioral Neurosciences	Henry Ford Hospital and Wayne State College of Medicine	Detroit, MI
Darrel Drobnich	Senior Director, Governmental Affairs	National Sleep Foundation	Washington, DC
Walter Edwards	Research Department	International Brotherhood of Teamsters (IBT)	Washington, DC
June Fisher	Director, MUNI Health & Safety Project and Associate Clinical Professor of Medicine	University of California, San Francisco	San Francisco, CA
Eric Garshick	Assistant Professor	Harvard Medical School	West Roxbury, MA
Keri Gentilcore	Director, Workplace and Fleet Safety	American Trucking Associations	Alexandria, VA
Janie Gittleman	Associate Director, Safety & Health Research	Center to Protect Workers Rights	Silver Spring, MD
Sarah Giuliani	Fitness Trainer/Coordinator	SGLLC/Roadwork	Grand Rapids, MI
Lonnie Golden	Associate Professor of Economics	Pennsylvania State University, Abington	Abington, PA
Patrick Hamelin	Director of Research	National Research Institute for Transport Safety (INRETS)	Arcueil, France
Karen Heaton	Doctoral student, Nursing	University of Kentucky	Louisville, KY

| \multicolumn{4}{c}{Participants, April 24–25, 2003, Conference on Truck Driver Occupational Safety and Health} |

Name	Title	Institution	City
Edward Hitchcock	Senior Service Fellow, Human Factors and Ergonomics	National Institute for Occupational Safety and Health	Cincinnati, OH
Lee Husting	Chief, Intervention & Evaluation Section and now Health Scientist / Program Administrator	National Institute for Occupational Safety and Health and Centers for Disease Control and Prevention	Morgantown, WV; Atlanta, GA
Ron Jager	Owner-operator (truck driver)	Ro-Jan	Taylor, MI
David Kameras	Communication coordinator	Teamsters Union	Washington, DC
Gerald Krueger[1]	Principal Scientist/Ergonomist	The Wexford Group International	Vienna, VA
Brenda Lantz	Program Director, Transportation Safety Systems Center	NDSU Upper Great Plains Transportation Institute	Lakewood, CO
Scott Madar[2]	Assistant Director, Safety and Health Department	Teamsters Union	Washington, DC
Michael McCann	Director of Safety & Ergonomics	Center to Protect Workers' Rights	Silver Spring, MD
James McGlothlin	Associate Professor of Health Sciences	Purdue University	West Lafayette, IN
Penelope Moyers	Dean, School of Occupational Therapy	University of Indianapolis	Indianapolis, IN
Audrey Newell	Associate Director, Family Practice Education Program	Oakwood Healthcare System	Dearborn, MI
Katharine Newman	Division Chief, Safety & Health Program Analysis & Control	Bureau of Labor Statistics	Washington, DC

---

[1] Prepared presentation for the conference, but was unable to attend.
[2] Currently a consultant with ORC Worldwide

\multicolumn{4}{c}{**Participants, April 24–25, 2003, Conference on Truck Driver Occupational Safety and Health**}			
Name	Title	Institution	City
Arthur Oleinick	Associate Professor, Environmental Health Sciences	University of Michigan	Ann Arbor, MI
Thomas Perrot	Senior Project Manager	ANTARES Group	Landover, MD
Peter Phillips	Professor of Economics	University of Utah	Salt Lake City, UT
Dieter Plehwe	Research Scholar at WZB and at the time of the conference, Lecturer, Department of Sociology	Social Science Center Berlin and Yale University	Berlin, Germany; New Haven, CT
Len Poirier	President, Local 4268	Canadian Auto Workers	Waterford, ON
Stephanie Pratt	Epidemiologist, Division of Safety Research	National Institute for Occupational Safety and Health	Morgantown, WV
Mark Price	Student, Department of Economics	University of Utah	Salt Lake City, UT
Les Pryce	Financial Secretary, Local 4268	Canadian Auto Workers	Scarborough, ON
Michael Quinlan	Professor of Industrial Relations and Organisational Behaviour	University of New South Wales	Sydney, Australia
Kathryn Reid	Research Assistant Professor, Center for Sleep and Circadian Biology	Northwestern University	Evanston, IL
Daniel Rodriguez	Assistant Professor and now Associate Professor, City and Regional Planning	University of North Carolina	Chapel Hill, NC
William Rogers	Vice President, Safety, Training, and Technology	Motor Freight Carriers Association	Washington, DC

| \multicolumn{4}{c}{**Participants, April 24–25, 2003, Conference on Truck Driver Occupational Safety and Health**} |

Name	Title	Institution	City
Lawrence Root	Professor and Director, Institute of Labor and Industrial Relations	University of Michigan	Ann Arbor, MI
Roger Rosa	Senior Scientist, Office of the Director	National Institute for Occupational Safety and Health	Washington, DC
Robert Rothstein	Vice President and General Counsel	American Moving and Storage Association	Alexandria, VA
Gregory Saltzman	Professor of Economics and Management	Albion College	Albion, MI
Ian Savage	College Lecturer and Assistant Department Chair, Economics	Northwestern University	Evanston, IL
Nancy Scibek	Research Division	Federal Motor Carrier Safety Administration	Washington, DC
John Sestito	Assistant Director for Surveillance	National Institute for Occupational Safety and Health	Cincinnati, OH
John Siebert	Project Manager	Owner-Operator Independent Drivers Association Foundation	Grain Valley, MO
Bethany Slingerland	Government Affairs and Transportation	National Sleep Foundation	Washington, DC
Thomas Smith	Professor of Environmental Health	Harvard School of Public Health	Boston, MA
David Snyder (participated by conference call)	Vice President and General Counsel	American Insurance Association	Washington, DC
Jeffrey Stern	Professor of Psychology	University of Michigan-Dearborn	Dearborn, MI

| \multicolumn{4}{c}{**Participants, April 24–25, 2003, Conference on Truck Driver Occupational Safety and Health**} |

Name	Title	Institution	City
Meg Sweeney	Office of Advanced Studies	Bureau of Transportation Statistics	Washington, DC
Wendine Thompson-Dawson	Doctoral student, Economics	University of Utah	Salt Lake City, UT
Hans Van Dongen	Research Assistant Professor of Sleep and Chronobiology and now Associate Research Professor	University of Pennsylvania School of Medicine; Washington State University	Philadelphia, PA; Spokane, WA
Patricia Waller (Prepared presentation for conference but unable to attend; now deceased)	Senior Research Scientist, Center for Transportation Research; Director Emerita, University of Michigan Transportation Research Institute	Texas A&M University; University of Michigan Transportation Research Institute	Chapel Hill, NC
Ann Williamson	Director, NSW Injury Risk Management Research Center	University of New South Wales	Sydney, Australia

Speaker Profiles

Torbjorn Akerstedt, PhD, is Professor of Behavioral Physiology at the Karolinska Institute. He is the past president of the European Sleep Research Society and recipient of the University of Pisa Sleep Award. He has authored 100+ papers on sleep, stress, and work hours. He is the Associate Editor of the Journal of Sleep Research and Biological Psychology, and is on the editorial board of several other journals. He has organized several international meetings on sleepiness and transport.

Michael H. Belzer, PhD, is Associate Professor of Industrial Relations at Wayne State University and Adjunct Associate Research Scientist at the University of Michigan's Institute of Labor and Industrial Relations. Dr. Belzer's research interests include all facets of trucking industry organization and operations, labor-management relations, and employment policy. He has studied the relationship between truck driver pay (both method and level) and safety, as well as issues related to truck driver hours of work. Dr.

Belzer currently serves as the Associate Director of the Sloan Foundation's Trucking Industry Program (TIP). He is Director of the Wayne State University Trucking Industry Benchmarking Program. He is the founding chairman of Transportation Research Board's Task Force on Trucking Industry Research and currently serves as Chairman of the Committee on Trucking Industry Research of the Transportation Research Board and is also a member of the TRB Freight Systems Executive Board. He is a member of the TRB Truck and Bus Safety Committee and also serves as a member of the Committee on Freight Regulation and Economics, and served on the Steering Committee for the 2005 National Safety Council Truck and Bus Safety and Security Symposium. From 2000 to 2003, Dr. Belzer served on the Committee for Review of the Federal Motor Carrier Safety Administration's Large Truck Crash Causation Study, formed by Transportation Research Board under authority granted by Congress to the National Research Council. He is the author of *Sweatshops on Wheels* as well as numerous peer-reviewed articles on industrial relations with special attention on the trucking industry.

Stephen V. Burks, PhD, is an Associate Professor of Economics and Management at the University of Minnesota, Morris. His B.A. is from Reed College, and he has an M.A. from Indiana University, Bloomington, in philosophy, focusing on the philosophy of science. He spent a decade working at blue-collar jobs in the trucking industry, during the period surrounding deregulation, after which he returned to graduate work at the University of Massachusetts Amherst, in economics. He received his Ph.D. in 1999, writing a dissertation on how the microeconomics of reciprocity and equity, in the employment relationship, affected the restructuring of the labor and product markets in U.S. motor freight after deregulation. While writing, he worked as a graduate researcher for the University of Michigan Trucking Industry Program (UMTIP, a Sloan Foundation program), with whom he then did a post-doctoral fellowship. He is pursuing two overlapping research directions; the first on the role of social preferences (such as trust and reciprocity) in employment relationships, using behavioral economic experiments framed in field settings (currently supported by the University of Minnesota and the MacArthur Foundation), and the second on several aspects of trucking industry economics (supported in 2000 and 2001 by UMTIP, and in 2003 by the Trucking Industry Program, Georgia Institute of Technology).

Orfeu Marcello Buxton, PhD, is an instructor at Harvard Medical School in the Division of Sleep Medicine. After receiving a B.S. in Behavioral Neuroscience from the University of Pittsburgh and a stint as the owner of a fitness equipment repair and moving business, Dr. Buxton began his graduate work in Neuroscience at Northwestern University. In work with Dr. Fred W. Turek, he primarily examined the interaction of circadian rhythms and sleep in rodents. For his thesis work, in collaboration with Dr. Van Cauter, he studied the effects of exercise, sleep, darkness, and a hypnotic on the timing and adjustment of the human circadian system. Dr. Buxton's recent work examined the effects of high-intensity exercise on metabolism and neuroendocrine secretion that vary with the time of day of the exercise.

Ralph Craft, PhD, has worked in Washington over the past 22 years as a: Transportation Lobbyist for the National Conference of State Legislatures; Assistant Executive Director of the International Bridge, Tunnel and Turnpike Association; Director of Transportation

Research for the National Governors' Association; Transportation Lobbyist for the State of New Jersey; Transportation Specialist for the Office of Motor Carriers, Federal Highway Administration; and presently as a Crash Data Analyst for the Federal Motor Carrier Safety Administration.

In a previous life, Mr. Craft earned a Doctorate in Political Science from Rutgers University; worked for and with a number of State legislatures; and published a book and many articles about state legislative processes and procedures, including several on State legislative review and evaluation of State government programs.

Drew Dawson, PhD, is the Dean of Research at the University of South Australia in Adelaide and Director of the internationally recognized Centre for Sleep Research (CFSR). The CFSR is the largest dedicated sleep research facility in Australia. Dr. Dawson has gained critical acclaim for his work and is considered a leading international authority in the areas of sleep, fatigue, biological rhythms, and hours of work.

Dr. Dawson has worked extensively with key transport organizations in the rail, road, and aviation industries both in Australia and abroad. He currently heads the Australian Rail Industry Consortium, focused on the effects of fatigue on shift workers in the field. This endeavor has led to the integration of training and education materials for a wide range of industries such as mining, manufacturing, and service. Additionally, fatigue-modeling software has also been developed for assessing and predicting fatigue levels associated with irregular work schedules, allowing better design and management of work hours. Collaborating with Transport Canada, the US Federal Rail Administration and the North American Motor Carriage Industry, Dr. Dawson has helped develop policy for some of the largest transportation networks in the world.

Dr. Dawson has produced over 60 industry reports, published hundreds of articles and papers, and attracted more than 5 million dollars in research grants. Dr. Dawson has also been a consultant to over 100 major Australian companies such as BHP, Mobil Industries Australia, Qantas, and the Kimberly-Clark Corporation.

Christopher Drake, PhD, serves as a member of the Senior Bioscientific Staff at The Henry Ford Hospital Sleep Disorders and Research Center in Detroit, Michigan. He graduated with a degree in Clinical Psychology from Bowling Green State University. Previously he served as a fellow at the National Institute of Mental Health, Clinical Psychobiology Branch under Dr. Thomas Wehr. He has served as a reviewer for the Journal SLEEP, Psychophysiology, Psychological Assessment, Small Group Research and Ergonomics. Dr. Drake is currently funded to study the predisposition to transient sleep disturbance and the evolution to chronic primary insomnia. He has been the author of 30 publications in the fields of sleep medicine and clinical psychophysiology. His primary areas of interest include the pathophysiology of insomnia and the causes and consequences of daytime sleepiness.

June Fisher, MD, is currently a senior scientist at the Trauma Foundation, San Francisco General Hospital. Her early career was in internal medicine and basic research in immunology. She has done extensive research on the immunological aspects of Pernicious Anemia. In the early 1970's she worked as a primary care physician in a neighborhood health center and did health services research at Stanford University. Since 1978 she has practiced occupational medicine. For 12 years she headed the Center for Municipal Occupational Safety and Health at San Francisco General Hospital. The hospital's Employee Health Service was incorporated into this center.

She heads two programs: the Training for Development for Innovative Control Technology (TDICT) Project and the Muni Health and Safety Study. These programs reflect her research activities in areas of health and safety of health care workers and urban public transit operators. She has published extensively on occupational stress and hypertension in urban transit operators. The studies on transit worker health and safety are currently funded by the National Institute for Health. The research projects have been done in collaboration with the San Francisco Municipal Railway and Local 250A, Transport Workers Union.

She is a member of the NIOSH Occupational Disease NORA Task Force. She was the co-organizer of the first international meeting on urban transit in Stockholm in 1983 and has organized several other international panels - the latest one was at the March 2003 occupational stress meeting sponsored by the American Psychological Association and the National Institute for Occupational health and Safety.

She was the editor for the Urban Transit Bulletin, which was published by the International Transport Workers Federation.

Eric Garshick, MD, is an Associate Professor of Medicine at the Harvard Medical School. He is also a Staff Physician at the VA Boston Healthcare System and a Physician, Research Staff, Channing Laboratory, Department of Medicine, Brigham and Women's Hospital. He received an MD in 1979 from Tufts University School of Medicine, an MOH from the Harvard School of Public Health in 1984, and is Board Certified in Internal Medicine, Pulmonary Disease, and Critical Care Medicine. He has been a consultant of the US EPA Science Advisory Board, Clean Air Scientific Committee for Diesel Emissions. His research interests are the health effects of diesel exhaust exposure and the epidemiology of chronic lung disease. He is currently Principal Investigator on National Cancer Institute grants studying the relationship between lung cancer and diesel exhaust exposure in railroad workers and trucking company workers.

Janie Gittleman, PhD, is Associate Director for Safety and Health Research at the Center to Protect Workers' Rights. She received her MRP and PhD from Cornell University in Health Planning and Epidemiology. Dr. Gittleman joined the National Institute for Occupational Safety and Health in 1990 as an Epidemic Intelligence Service Officer, and held positions as Section and Branch Chief of the Surveillance and Hearing Loss Prevention Branches, in Cincinnati, Pittsburgh, and Atlanta from 1992-1998. From 1999-2001, she served as a Senior Scientist in the NIOSH Office of the Director in Washington, D.C. Dr. Gittleman previously served on the faculty of the Medical College of Pennsylvania in the Department of Preventive Medicine.

Dr. Gittleman specializes in occupational disease surveillance. She established the Adult Blood Lead Epidemiology and Surveillance Program at NIOSH, which currently monitors adult lead poisoning in 35 states nationwide. She has served on numerous CDC occupational disease surveillance committees, and was hired by the United States Agency for International Development, the Pan American Health Organization, and the World Bank to develop surveillance for the reduction of lead in gasoline in Latin American and the Caribbean. She recently served as staff member to the Secretary of Health in the newly created Office of Public Health Preparedness, and developed benchmarks for state accountability for emergency response planning for bioterrorism. She is the author of numerous scholarly publications in occupational safety and health with an emphasis on state-based disease surveillance.

Patrick Hamelin, PhD, is a Research Director at the French National Institute for Transport and Safety Research (INRETS). He studied Sociology and Economics at Paris University "La Sorbonne", and developed his social science research skills at the Ecole Pratique des Hautes Etudes en Sciences Sociales. His previous work includes research on the effects of changes in agricultural production on rural space and ways of life, the organization of urban transport and the growth patterns of towns, and work organization in transport. His research focuses on the relationships among work organization and work environment, conditions of work, time of work, kind of wages, safety, and the social properties of truck drivers. He has published several articles and books in French and in English.

Dr. Hamelin's work was quoted in Gearing up for Safety, a report published in 1988 by the U.S. Congress, Office of Technology Assessment. This report called for increased attention to driver hours of service and fatigue and safety.

Gerald P. Krueger, PhD, has over 36 years experience as an engineering psychologist and ergonomist conducting, managing, and directing multidisciplinary research and applications in measurement and prediction of worker performance. He specializes in the human performance implications of equipment operator fatigue, sleep deprivation, and sustained operations. Dr. Krueger obtained his Ph.D. in engineering and experimental psychology from the Johns Hopkins University. He is a graduate of the US Army War College and is a certified professional ergonomist (CPE). He is a retired Army colonel who served 25-years in Army R&D, culminating in his tour as the military commander and technical director of the U.S. Army Research Institute of Environmental Medicine (USARIEM).

Since Army retirement, Dr. Krueger has spent seven years examining commercial driver fatigue, and assessing wellness, fitness and driver alertness. In his work for the US DOT Federal Motor Carrier Safety Administration's (FMCSA) and the American Trucking Associations, he is the principal lecturer on driver alertness, fatigue, and wellness, health and fitness; and he is the operations manager of a large on-the-road field experiment evaluating fatigue management technologies with truck drivers in Canada and in the US.

Dr. Krueger has authored over 80 technical publications, book chapters, and contract reports – many of them on operator alertness and fatigue – and he continues to make numerous formal speeches and presentations at national and international scientific meetings and conferences. Dr. Krueger fulfills senior human factors consultant roles for several federal agencies, including Dept. of Defense Research and Engineering, the Defense Information Systems Agency, the Nuclear Regulatory Commission, and the Department of Transportation. He is a Fellow and is very active in two divisions of the American Psychological Association (Military Psychology and Applied Experimental and Engineering Psychology); is active in technical journal editing, and in leadership roles in other Human Factors and Ergonomics Society positions.

Brenda Lantz, PhD, is the Program Director for the Transportation Safety Systems Center branch of the Upper Great Plains Transportation Institute. The Center is responsible for software development for commercial vehicle safety enforcement programs, as well as safety-related research and analysis. Brenda holds a Masters degree in Statistics from the North Dakota State University, and a Ph.D. in Business Logistics from Penn State. She has more than 12 years of experience in transportation research, primarily in the commercial vehicle safety systems field, and has worked extensively with both government and private industry agencies.

Scott A. Madar, CIH, is a consultant with ORC Worldwide. Previously, Mr. Madar has worked as an OSHA compliance officer in North Carolina and spent over 10 years working with the International Brotherhood of Teamsters. While serving as the Assistant Director of the Safety and Health Department of the IBT, Mr. Madar had the opportunity to testify before three congressional subcommittees and the National Transportation Safety Board on a variety of transportation safety issues. He has also served as the chair of the Occupational Safety and Health Subcommittee of the Labor Research Advisory Committee for the Bureau of Labor Statistics. He currently serves on the Commercial Truck and Bus Safety Synthesis Program of the Transportation Research Board. Mr. Madar recently organized a National Summit on Contractor Safety on behalf of the Duke Energy Foundation.

After Mr. Madar obtained his B.S. in biology from The College of William and Mary, he went on to earn his M.H.S. in industrial hygiene and safety sciences from The Johns Hopkins University. He is a Certified Industrial Hygienist (CIH).

Michael McCann, PhD, CIH, is Director of Safety and Ergonomics at the Center to Protect Workers' Rights, the research arm of the Building and Construction Trades Department, AFL-CIO. He received his PhD in Chemistry from Columbia University in 1972, became a certified industrial hygienist in 1979, and is authorized to conduct 10- and 30-hour OSHA courses as an OSHA Construction Outreach Trainer. Dr. McCann participated in a study of hazards of Ready Mixed concrete truck drivers. He has also been studying the hazards of heavy equipment such as bulldozers and dump trucks on construction excavation sites. He has given presentations at the Construction Safety Council Annual Conference and other construction safety conferences, and at safety professional conferences. He helped develop and conduct training for construction workers, including dump truck drivers, at the World Trade Center Emergency Project in 2001.

James D. McGlothlin, MPH, PhD, CPE, is an Associate Professor of Industrial Hygiene and Ergonomics in the School of Health Sciences at Purdue University. Dr. McGlothlin specializes in research in ergonomics and in industrial hygiene engineering controls. Prior to Dr. McGlothlin's appointment to Purdue University, (January 4, 1999) he was a senior researcher in ergonomics with the Engineering Control Technology Branch, Division of Physical Sciences and Engineering, National Institute for Occupational Safety and Health (NIOSH), Cincinnati, Ohio. The author of more than 100 scientific, technical, and government reports, Dr. McGlothlin served as a course director in ergonomics at Northwestern University School of Engineering, Evanston, Illinois, and at the University of Cincinnati School of Medicine, Ohio. He has received several awards for his research and service to the U.S. Public Health Service, including the Surgeon General's Exemplary Service Medal, the Outstanding Service medal, and Commendation Medal. He has a Patent (6,094,780) on an ergonomically designed handle. He has served on several national and international professional committees, and currently serves on the editorial boards of the Occupational Hazards Journal. Dr. McGlothlin received the B.A. degree (1975) in psychology, the M.P.H. degree (1977) in epidemiology, and the M.S. degree (1977) in environmental and industrial health, all from the University of Hawaii, Honolulu. He received the Ph.D. degree (1988) in industrial health with a specialty in ergonomics from the University of Michigan, Ann Arbor. Dr. McGlothlin is a Certified Professional Ergonomist. You can find out more about Dr. McGlothlin by visiting his website at *www.DrMcGlothlin.com*.

Katharine Newman is an Economist and the Acting Assistant Commissioner for Occupational Safety, Health, and Working Conditions at the Bureau of Labor Statistics. She has been with the BLS for 11 years, joining the occupational safety and health statistics programs in 1999. In her current position, Ms. Newman has responsibility for all aspects of the Annual Survey of Occupational Injuries and Illnesses and the Census of Fatal Occupation Injuries. Previously, she worked for the U.S. Bureau of the Census, the Office of Management and Budget, and the Brookings Institution.

Stephanie Pratt, MM, MA, is an epidemiologist at the National Institute for Occupational Safety and Health, Division of Safety Research, in Morgantown, West Virginia. Her major research interest is occupational motor vehicle safety. She is the author of the forthcoming NIOSH document "Work-related Roadway Crashes: Challenges and Opportunities for Prevention," as well as the 2001 NIOSH document "Building Safer Highway Work Zones." Ms. Pratt has also published papers on adolescent workers, farm fatalities of youth, older workers, and machinery-related fatalities. She contributes regularly to NIOSH policy responses to notices of proposed rulemaking. Before coming to NIOSH in 1993, Ms. Pratt worked in the West Virginia University, where she managed the State Census Data Center and was involved in research projects on maternal and child health. Ms. Pratt holds master's degrees from West Virginia University in applied music and applied social research.

Michael Quinlan, PhD, MSIA, has been involved in occupational health and safety (OHS) education, research, practice and policy advice for over twenty years. He is the author of numerous articles on occupational health and safety as well as the author or editor (including joint editor) of a number of books, including *Managing Occupational Health and Safety* (Macmillan 1991 and 2nd edition 2000), *Work and Health* (Macmillan 1993) and *Systematic OHS Management* (Elsevier, Oxford, 2000). He has acted as a policy adviser to both State and Federal governments in Australia on OHS for over ten years, serving on a number of committees and working parties. In 2000/2001, he undertook an inquiry into safety in the Australian long haul trucking industry for the Motor Accidents Authority of New South Wales. He is currently undertaking a study of OHS amongst short haul truck drivers as part of a broader project on contingent work arrangements. His particular interests include management systems/work organization, outsourcing/contingent work and the legal regulation of OHS. Apart from invitations to give papers at international conferences and workshops, he is a regular speaker at industry conferences and forums.

Kathryn J. Reid, PhD, received her PhD from the University of Adelaide (Australia) in 1998. She is currently a Research Assistant Professor at Northwestern University in the Center for Sleep and Circadian Biology and the Transportation Center. Dr Reid has worked closely with the rail industry in Australia investigating the impact of irregular hours of work on sleep, performance, and safety. Her current research interests are on the impact of sleep loss and circadian disruption on sleep, performance, health, and safety. She is currently developing a new curriculum on drowsy driving for driver education programs in Illinois, in collaboration with the Illinois Department of Public Health. She has published her work in Nature and in other peer-reviewed journals, and has presented at numerous conferences and meetings.

Daniel A. Rodríguez, PhD, is Associate Professor of City and Regional Planning, UNC-Chapel Hill BS, 1994, Fordham University; Master of Science in Transportation, 1996, Massachusetts Institute of Technology, and Ph.D. in Planning, 2000, University of Michigan. Dr. Rodriguez teaches graduate courses in transportation policy, transit planning, and urban spatial structure. His research includes behavioral aspects of travel demand and truck transportation policy. He has conducted research on transportation safety, public transportation, multi-objective decision-making, and intelligent transportation systems. Dr. Rodríguez is currently appointed to three standing committees of the National Academies' on Transportation and Land Development, Travel Demand Management, and Freight Economics and Regulation, and to the Task Force on Trucking Industry Research. Professor Rodríguez has been the recipient of the Transportation Research Board's Fred Burggraff Award (2000) and the Eno Foundation Fellowship (1998). He has worked internationally for the World Bank and the Universidad de los Andes, Bogotá, Colombia. Dr. Rodríguez is also an Eno Transportation Foundation Leadership Fellow and a Fellow at NYU's Institute for Civil Infrastructure Systems.

Dr. William C Rogers, PhD, has over 20 years of experience in a wide variety of safety, training, and research endeavors in the trucking industry. He is Vice President of Safety, Training & Technology at the Motor Freight Carriers Association, a non-profit trade association established to advance the economic interests of the unionized, less-than-truckload motor carriers. He was the project manager for a recently completed study of 22,000 workplace injuries and is currently conducting a survey of Teamster truck drivers with one million accident free driving miles to determine the techniques they use to manage fatigue, particularly during night operations. While serving as Director of Research for the ATA Foundation (the research and education arm of the American Trucking Associations) he defined the research needs of the trucking industry, identified funding sources, and managed and conducted trucking research for government and industry sponsors. His projects included research on the adequacy of truck parking at public rest areas, a comprehensive fatigue education program for truck drivers, the truck driver shortage, the prevalence of sleep apnea among truck drivers, and workplace injuries in the trucking industry. He also served in several positions as a civilian employee with the US Army, including, Safety director of the US Army Transportation Center, Safety Manager at Headquarters, US Army Europe, and Safety Manager of the 37th Transportation Group, the Army's line haul motor carrier in Europe. He earned his Doctorate in Safety Education at Texas A&M University.

Lawrence S. Root, PhD, is director of the University of Michigan Institute of Labor and Industrial Relations and a professor in the School of Social Work. His area of research is the interplay between employment and social well-being. He has worked extensively with joint union-management programs. For example, he directed a major educational joint counseling program, which provided educational assistance to hourly workers in plants throughout the U.S. He also directed a distance learning college program specifically designed for autoworkers. For several years, Professor Root has served as chairman of the University's Committee on Labor Standards and Human Rights, which is advisory to the President of the university on issues related to the manufacture of licensed goods. He is also the director of the Labor and Global Change program, which explores issues relevant to labor conditions in a cross-national context.

Roger Rosa, PhD, is currently a Senior Scientist in the Office of the Director of NIOSH in Washington DC. He began his service with NIOSH in 1984 as a Research Psychologist in the Cincinnati laboratories following completion of a Ph.D. in Experimental Psychology from the University of Cincinnati. Throughout his career, Dr. Rosa has performed research on human performance and physiology as affected by shift work, biological rhythms, sleep loss, and fatigue. He has authored numerous articles and edited books on these and other occupational health and safety topics, served on national research committees, and collaborated with other agencies on issues related to working hours, stress, and fatigue.

Robert G. Rothstein, JD, is Vice President and General Counsel of the American Moving and Storage Association ("AMSA") in Alexandria, Virginia, where he specializes in legal, regulatory, and safety matters related to the moving and storage segment of the motor carrier industry.

Mr. Rothstein has been with AMSA since October 2001. He spent nine years as general counsel with the Truckload Carriers Association, and was previously engaged in the private practice of law, concentrating in labor and employment law and litigation on behalf of management. He also spent seven years as a staff attorney with the Interstate Commerce Commission.

Mr. Rothstein holds a B.B.A. degree in Finance from the University of Miami, a J.D. from Nova University, and a Master of Laws degree in labor law from Georgetown University. Between college and law school he hauled frozen vegetables between Florida and the Northeast. He maintains a current commercial driver's license (CDL).

Gregory M. Saltzman, PhD, is E. Maynard Aris Professor of economics and management at Albion College and adjunct research scientist at the Institute of Labor and Industrial Relations, University of Michigan. Previously, he taught social policy at Brandeis University and labor relations at Ohio State University. A paper by Greg Saltzman and Mike Belzer, "The Case for Strengthened Motor Carrier Hours of Service Regulations," was published in the Summer 2002 issue of *Transportation Journal*. This paper summarized the medical and epidemiological literature on the impact of long and irregular work schedules on driver health, identified market failures that justify government regulation of driver hours, and noted statutory requirements that the U.S. Department of Transportation protect driver health. Dr. Saltzman's other papers include studies of labor law, personnel selection procedures, reimbursement systems for mental health services, public-sector collective bargaining, and union organizing. He received his training as an economist at MIT, the London School of Economics, and the University of Wisconsin–Madison.

Ian Savage, PhD, has been a member of the faculty of both the Department of Economics and the Transportation Center at Northwestern University since 1986. He is also the Assistant Chairperson of the Economics Department, and during 2000-01 was the Acting Associate Director of the Transportation Center. He earned a bachelor's degree in economics from the University of Sheffield and a Ph.D. from the School of Economic Studies/Institute for Transport Studies at the University of Leeds. Dr. Savage specializes in urban transportation, and the analysis of safety regulation and safety performance. He has conducted research into the safety of most modes of transportation. Most recently, he completed a book on *The Economics of Railroad Safety* (Kluwer Academic Publishers, 1998). Prior to that, he was part of a five-year study of the economics of safety regulation in the trucking industry. He has also published widely on the economics of transit finances and operations and, more specifically, the impacts of competition and privatization.

Mr. John Siebert, MS, EdSp, is a researcher and communications/training specialist, who works in the OOIDA Foundation producing educational materials and training modules specifically for independent truckers. He also performs primary survey research concerning driver behavior and attitudes. He has a lengthy background in academic and corporate communications and training. He earned his B.S., M.S., and Ed.Sp. degrees from Central Missouri State University.

Thomas J. Smith, PhD, CIH, is currently Professor and Director of the Industrial Hygiene Program at the Harvard School of Public Health. Dr. Smith has long been interested in the design and implementation of exposure assessments for epidemiologic studies, particularly the estimation of past exposures. Recently a new assessment was begun as part of an NCI funded cohort study with Dr. Eric Garshick, Harvard Medical School, which will investigate lung cancer mortality in the US trucking industry. This new study will attempt to determine if there is a distinct risk from diesel engine emissions, as opposed to general combustion emissions and particles from all sources, such as trucks, automobiles, and home heating. Dr. Smith is also interested in the relationship between external exposure and internal dose, specifically the uptake and metabolism processes. He and his colleagues have developed a computer controlled apparatus to briefly expose volunteers to low levels of airborne chemicals (doses in the range of every day exposures) while making precisely timed measurements of exhaled breath levels during and after exposure. These data are used to estimate each subject's uptake and metabolism of the material, which can better define their personal dose and risk from the chemical and who has the highest risk. All of Dr. Smith's graduate work was performed at the University of Minnesota.

David F. Snyder, JD, is Vice President and Assistant General Counsel for the American Insurance Association (AIA), specializing in legal reform, automobile insurance and other transportation issues. He also provides staff support for AIA's International Committee. His responsibilities include drafting legislation and regulations, testifying, managing litigation and communicating with the media.

Mr. Snyder joined AIA in 1987 after working for Nationwide Insurance, managing its government affairs in its largest premium volume state. He left AIA in 1990 to work for State Farm Insurance Company and later returned to AIA. Prior to these positions, he served the Commonwealth of Pennsylvania as General Counsel of the Commerce Department, Deputy Attorney General in Torts Litigation and Legislative Liaison for the Insurance Department. Earlier, he was counsel to a federal agency. These positions included trial and appellate litigation, in-house legal counsel and government affairs responsibilities.

Meg Sweeney, PhD, is a project manager/technical report writer for the Office of Highway Safety at the National Transportation Safety Board. She returned to the Safety Board after serving two years as a Transportation Specialist at the Bureau of Transportation Statistics. From 1994 to 2002, Dr. Sweeney was a Transportation Research Analyst in the Safety Studies Division in the Office of Research and Engineering. Prior to her service at the NTSB, she was a Project Manager at Advanced Resource Technologies, Inc. and a Consortium Research Fellow at the U.S. Army Research Institute. Dr. Sweeney received her B.A. in Psychology from Boston College and her M.A. and Ph.D. in Applied Experimental Psychology from George Mason University.

Hans P.A. Van Dongen, PhD, is Research Assistant Professor of Sleep and Chronobiology in the Department of Psychiatry, University of Pennsylvania School of

Medicine. He received an M.S. degree in astrophysics and a Ph.D. degree in chronobiology and sleep from Leiden University in the Netherlands. The focus of his work is on inter-individual variability in sleep and wakefulness; previous and current projects investigated inter-individual differences in circadian phase and inter-individual differences in vulnerability to performance impairment from sleep loss. Dr. Van Dongen recently published a study of the neurobehavioral effects of chronic sleep restriction relative to acute total sleep deprivation with David F. Dinges. The results led to the formulation of a novel "cumulative excess wakefulness" hypothesis for progressive impairments from sleep loss. Dr. Van Dongen is also active in the development of statistical techniques for time series analysis, which serves as a basis for his involvement in the biomathematical predictive modeling of performance deficits from sleep loss for the prevention of sleepiness-related human error.

Patricia F. Waller, PhD, was a Senior Research Scientist with the Center for Transportation Safety at Texas A & M University. She previously served as Director of the Transportation Research Institute at the University of Michigan, retiring in 1999. Prior to going to Michigan in 1989, she was Associate Director for Driver Studies at the University of North Carolina Highway Safety Research Center, Research Professor in the UNC School of Public Health, and founding director of the UNC Injury Prevention Research Center. She served on the Board of Directors of ITS AMERICA from 1991 through 1999 and chaired the Transportation Research Board's Group 3 Council, Operations, Safety, and Maintenance of Transportation Facilities, and the Group 5 Council, Intergroup Resources and Issues.

Her background in heavy truck transportation began in the early 1970's. She worked with the U.S. DOT, including FHWA, Office of Motor Carriers (BMCS previously), and NHTSA on heavy truck issues. In 1976 she served on a 6-member TRB panel to review and advise FHWA on their Heavy Truck Safety 5-Year Research Plan. She has served as Principal Investigator on several federal contracts concerning heavy truck safety, and served on the TRB Committee on Motor Vehicle Size and Weight (heavy trucks) since it began as a task force in 1979 to 1999. She also was a member of the TRB Special Committee for Truck Safety Data Needs Study; and on the Human Factors Workshop on Heavy Truck Safety for the Office of Technology Assessment. She served as Vice-Chair of the National Motor Carrier Advisory Committee and was subsequently appointed Chair. In 1995, she chaired the research sessions at the First National Conference on Truck and Bus Safety and was a participant in preparing the output of that meeting. She also worked with the Office of Motor Carrier Safety to develop guidelines, based on the input from the conference, to promote heavy truck safety through public and private initiatives. She continued to work closely with Office of Motor Carriers on such issues as driver fatigue, Hours of Service, and truck safety research needs. She passed away August 15, 2003.

Ann Williamson, PhD, conducts research in occupational health and safety. After working as a Research Scientist for the New South Wales Government in the Department of Industrial Relations, Division of Occupational Health for seven years, she moved to the National Institute of Occupational Health and Safety where she was Principal

Research Scientist and Head of the Human Factors and Ergonomics Unit. Ann's main research interests include the effects of fatigue and long and irregular working hours especially relating to long distance road transport, the role of behavior in the causes of injury and the neurobehavioral effects of exposure to workplace hazards. This work has continued following her move to the University of New South Wales, where Ann was founding Executive Director of the NSW Injury Risk Management Research Centre. Ann has published extensively in the scientific literature and has also been an invited speaker at a wide range of national and international conferences, and an invited member of a number of government committees on occupational health and safety.

Appendix 2: Agenda

Agenda, April 24–25, 2003, Conference on Truck Driver Occupational Safety and Health

Thursday, April 24, morning

Activity	Time	Topic	Speakers
Continental breakfast	7:30-8:30		
Plenary	8:30-10:30	An Introduction to Trucking	Dr. Michael Belzer
	8:30-8:40	Introduction and Purpose of the Conference	Dr. Michael Belzer
	8:40-8:45	Keynote: The Need for a Agenda on Truck Driver Occupational Safety and Health	Dr. Patricia Waller [unable to attend]
	8:45-8:55	How Many Trucks, How Many Miles? Heavy Freight Vehicles in the U.S., 1977-1997	Dr. Stephen Burks
	8:55-9:15	Sweatshops on Wheels: Winners and Losers in Trucking Deregulation	Dr. Michael Belzer
	9:15-9:35	Market Failures, Economic Efficiency, and the Need for Federal Regulation of Truck Driver Work Hours	Dr. Gregory Saltzman
	9:35-9:55	European Trucking: Health and Safety versus Competitiveness?	Dr. Dieter Plehwe
	9:55-10:15	Trucking in Australia: Lessons from the Recent Inquiry (2001) into OHS in Long Haul Trucking	Dr. Michael Quinlan
	10:15-10:30	Discussion	
Break	10:30-10:45		
Plenary	10:45-12:25	Surveillance	Chair: Dr. Janie Gittleman
	10:45-11:05	Fatigue and Sleepiness in the Workplace: Can We Track It?	Dr. Roger Rosa
	11:05-11:25	Truck Driver Injury and Illness Data from the Bureau of Labor Statistics	Katharine Newman
	11:25-11:45	Truck Crash Data: Truck Drivers	Dr. Ralph Craft
	11:45-11:55	Data Sources Available at BTS	Dr. Meg Sweeney
	11:55-12:05	An Analysis of CMV Driver Traffic Conviction Data to Identify High Safety Risk Motor Carriers	Brenda Lantz
	12:05-12:25	CFOI and FARS Data on Truck Crashes: Apples and Oranges?	Dr. Stephanie Pratt

Agenda, April 24-25, 2003, Conference on Truck Driver Occupational Safety and Health (continued)

Thursday, April 24, afternoon

Lunch	12:25-1:10	Welcome from the Dean of the College of Urban, Labor and Metropolitan Affairs Wayne State University and Lunch	Dean Alma Young CULMA
Plenary	1:10-2:55	Exposure, Injury, and Fitness	Chair: Dr. William Rogers
	1:10-1:30	Diesel Exhaust Health Effects: Trucking Industry Particle Study	Dr. Eric Garshick
	1:30-1:50	Workers Compensation Claims Experience in Unionized LTL Carriers	Dr. William Rogers
	1:50-2:10	An Ergonomic and Cardiovascular Stress Evaluation of Soft Drink Beverage Deliverymen	Dr. James McGlothlin
	2:10-2:25	Occupational Hazards of Ready Mixed Concrete Truck Drivers	Dr. Michael McCann
		Gettin' in Gear: Wellness, Health & Fitness for Commercial Drivers	Dr. Gerald P. Krueger [unable to attend]
	2:25-2:45	Discussion	
Break	2:45-3:00		
Plenary	3:00-5:15	The Economics of Safety	Chair: Dr. Michael Belzer
	3:00-3:30	Pay, Work Hours, and Driver Safety	Dr. Daniel Rodriguez
	3:30-3:50	Transportation Safety	Dr. Ian Savage
	3:50-4:20	Occupational Safety and Health in Trucking: An Australian and European Perspective	Dr. Michael Quinlan
	4:20-4:50	Professional Drivers' Working Time as a Factor of Flexibility and Competitiveness in Road Haulage	Dr. Patrick Hamelin
	4:50-5:30	Discussion	
Break	5:30-6:00		
Reception	6:00-6:30		
Dinner	6:30-7:30		
Panel Session	7:30-9:30	Panel Discussion: HOS regulations; reactions to presentations on health and safety; effects on industry, on driver work life and earnings	Robert Rothstein, Ron Jager, David Snyder [participated by conference call]

Agenda, April 24-25, 2003, Conference on Truck Driver Occupational Safety and Health (continued)

Friday, April 25

Continental breakfast	7:00-8:00		
Plenary	8:00-9:45	Safety Issues and Sleep	Chair: Scott Madar
	8:00-8:20	Effects of Fatigue on Safety: Tired or Drunk, What's the Difference?	Dr. Kathryn Reid
	8:20-8:40	Fatigue and Safety: Chronic Sleep Loss, Wake State Instability, and Inter-Individual Differences	Dr. Hans Van Dongen
	8:40-9:00	Chronic Sleep Loss Leading to Obesity, Diabetes	Dr. Orfeu Buxton
	9:00-9:20	Sleep Apnea: Prevalence, Risk Factors, & Consequences Relevant to the Trucking Industry	Dr. Chris Drake
	9:20-9:40	Pressure of Work Organization on Health: Lessons from Transit	Dr. June Fisher
	9:40-9:45	Discussion	
Break	9:45-10:00		
Plenary	10:00-12:00	Preliminary Analysis of OOIDA Member Data	Chair: John Siebert
	10:00-10:35	Work Hours, Sleep and Health	Dr. Torbjorn Akersted
	10:35-11:10	Evaluating Different Work-Rest Schedules and Arrangements for Long Distance Road Transport: Risk Factors for Fatigue and the Role of Regulation	Dr. Ann Williamson
	11:10-11:30	I was so Tired – I had a Vision: A Fatigue Risk Management System for Road Transport	Dr. Drew Dawson
	11:30-11:45	Work and Family Issues: The Effects of Long Work Hours	Dr. Lawrence Root
	11:45-12:00	Discussion	
Lunch	12:00-12:45		
Break outs	12:45-2:30	By area of interest	
Plenary	2:30-3:30	Review and Next Steps: Develop Research Agenda	Chair: Dr. Michael Belzer

www.ingramcontent.com/pod-product-compliance
Lightning Source LLC
Chambersburg PA
CBHW081727170526
45167CB00009B/3725